Developments in Health Economics and Public Policy

Volume 12

Series editors

H. Frech, Santa Barbara CA, USA
Peter Zweifel, Zurich, Switzerland

For further volumes:
http://www.springer.com/series/6039

Rosella Levaggi · Marcello Montefiori
Editors

Health Care Provision and Patient Mobility

Health Integration in the European Union

Lifelong
Learning
Programme

With the support of the Lifelong Learning Programme of the European Union

Springer

Editors
Rosella Levaggi
Department of Economics and Management
University of Brescia
Brescia
Italy

Marcello Montefiori
Department of Economics
University of Genoa
Genova
Italy

ISSN 0927-4987
ISBN 978-88-470-5869-9 ISBN 978-88-470-5480-6 (eBook)
DOI 10.1007/978-88-470-5480-6
Springer Milan Heidelberg New York Dordrecht London

Printed on acid-free paper

Springer is part of Springer Science+Business Media (www.springer.com)

Preface

Patient mobility is often described as a marginal phenomenon in Europe since the overall number of patients that receive cross border care remains minor compared to the total population. However this phenomenon is increasing markedly. The process of globalization and the availability of medical information means that patients are more informed on treatment available well beyond their national boundaries. Patient empowerment implies that new generations will actively ask to be treated by the health care system that best meets their needs. A succession of individuals has already challenged the status quo and, in many cases, the European Court of Justice has upheld their arguments. At a political level, the EU has issued the EU Directive no. 24/2011/CE of 9th March 2011 concerning the application of patients' rights in cross border healthcare. This Directive has reformed the authorization procedures that were previously required to allow patients to go abroad to access health care services. In this respect, the Directive has contributed to improving the level of freedom of choice for the European citizen, but it does not seem to have increased actual patient mobility across Europe. Freedom to choose is a necessary condition to grant the people of Europe the same access to public sector health care services. The latter is a key instrument for an efficiently functioning "single market" ensuring real mobility within the EU.

The aim of this book is to study the current European health care market and discuss the hypothesis of a European right of citizenship with reference to health care services. The book is intended to provide a deeper understanding of the health market and stimulate reflection on European integration studies at a university level. The hypothesis for a fully coordinated European Health System will be investigated in great detail, highlighting the connected social and economic implications.

This publication is a result of the Jean Monnet Lifelong Programme. Patient mobility for health within the EU borders is an issue of great interest in the European Union debate. The health side of European integration is generally undervalued but, on the contrary, it is crucial to fostering an effective internal market and to ensuring economic and social progress in Europe. Effective health integration in the European Union would represent a value-added resource in the path toward the effective mobility of people. For this purpose, it is extremely important that policy-makers of different European States agree on a unique international agreement to regulate the economic implications of patient mobility.

At present we can observe very different National Health Care Systems in Europe but the condition required for a full integration and the ability to cope with the increasing mobility of people inside the EU is a set of shared rules. The harmonization process is certainly very demanding, both from the economic and social point of view, but certainly it is a feasible target.

One of the goals of the present publication is to deepen the knowledge in advanced topics in the field of European integration in order to improve the awareness of the need for a common European health policy.

The current European health care market is considered in order to put forward the challenge of a European right of citizenship with reference to health. New viable solutions have to be identified and implemented to make it easier for people from the EU to access health care services everywhere in Europe even if very few citizens are aware of its relevance. The goal of a fully coordinated European Health System is still far from being achieved even though its existence is fundamental to granting people mobility.

The present book focuses on theoretical and empirical aspects concerning patient mobility and health integration, both from positive and normative perspectives. To this extent, topics such as health economics and health systems, health data collection and data analysis, health mobility flow, health funding systems and financial sustainability, common European indicators for quality and international transfers will be presented and discussed.

In the first chapter Brekke, Gravelle, Siciliani and Straume present the current state of affairs in cross border care.

Their review depicts what we know about patient mobility at a theoretical and empirical level. This chapter deals with actual choices of patients; the following chapter by Pfarr, Schmid and Schneider is instead an attempt to interpret where these choices come from through the study of patient preferences.

The empirical part presenting evidence on cross border mobility at EU and national level opens with the contribution by Kifmann and Wagner that presents German patients' choices concerning treatment abroad. The following chapter by Glinos studies the mobility of health care professionals within the EU, a very interesting and often neglected theme especially in view of the uneven distribution of some professional figures across Europe. Patient mobility at national level is studied using data for the UK and Italy.

Dusheiko presents the choices of UK patients regarding their hospital admission at national and international level and shows the responses of the National Health Service (NHS) to these choices in terms of rules to regulate the flow of patients.

Balia, Brau and Marrocu examine patient mobility across Italian regions using data on hospital discharges. What makes the Italian case more interesting is the decentralization of the NHS that yields to large regional variations in patient flows in favor of Center-Northern regions, which typically are 'net exporters' of hospital treatments. The empirical section of the book concludes with the contribution by Crivelli and Salari, which presents evidence of the impact of federalism and cross border shopping on the healthcare system in terms of efficiency, equity, and cost containment in the case of Switzerland. The Swiss system can, in fact, be used to draw important policy implications on the likely effects that health care integration may have at European level. The last three chapters of the book develop this line of analysis by studying the welfare implications of cross border shopping. Levaggi and Levaggi study the combined effects of restrictions on the use of health care, transfer prices, and mobility rules on social welfare. Montefiori focuses on hospital competition in a context where the latter might have asymmetric objectives and costs, which are reimbursed using a uniform prospective payment. Different equilibrium outcomes, under perfect and asymmetric information are provided.

Finally, Levaggi and Menoncin study cross border shopping in a context where equity issues are taken into account (through a super-national equalization grant) and health care is modeled as an impure public good.

This book provides new empirical and theoretical insights for a more comprehensive understanding of patient mobility and the health care market. It is intended for health researchers, decision-makers, and professionals concerned with health care provision and patient mobility. The goal is to provide, through scientific and methodological rigor, new informative tools useful for the implementation of new policies in the health care sector in order to implement effective health care integration in the European Union.

This project has been funded with support from the European Commission (Lifelong Learning Programme of the European Union). This publication reflects the views only of the authors and the Commission cannot be held responsible for any use which may be made of the information contained therein.

Brescia, Italy — Rosella Levaggi
Genoa, Italy — Marcello Montefiori

Contents

Patient Choice, Mobility and Competition Among Health Care Providers

Kurt R. Brekke, Hugh Gravelle, Luigi Siciliani and Odd Rune Straume

Abstract Policymakers are increasingly designing policies that encourage patient choice and therefore mobility across providers. Since prices are regulated (fixed) in most countries, providers need to compete on quality to attract patients. This chapter reviews the current theoretical and empirical literature on patient choice and quality competition in health markets. The theoretical literature identifies key factors affecting incentives to provide quality. These include: altruistic motives, cost structure, number of providers, demand responsiveness, GP gatekeeping, degree of specialization, profit constraints and soft budgets. We also review the theoretical literature on choice across different countries (e.g. within the EU) or regions within the same countries. The chapter reviews selected empirical studies that investigate whether demand responds to quality and waiting times, the role of patient's mobility and the effect of competition on quality.

Keywords Choice · Quality competition

K. R. Brekke
Department of Economics, Norwegian School of Economics, Helleveien 30 5045 Bergen, Norway
e-mail: kurt.brekke@nhh.no

H. Gravelle
Centre for Health Economics, University of York, Heslington, York, UK
e-mail: hugh.gravelle@york.ac.uk

L. Siciliani (✉)
Department of Economics and Related Studies, and Centre for Health Economics, University of York, Heslington, York, UK
e-mail: luigi.siciliani@york.ac.uk

O. R. Straume
Department of Economics/NIPE, University of Minho, Campus de Gualtar 4710-057 Braga, Portugal
e-mail: o.r.straume@eeg.uminho.pt

O. R. Straume
Department of Economics, University of Bergen, Bergen, Norway

R. Levaggi and M. Montefiori (eds.), *Health Care Provision and Patient Mobility*, Developments in Health Economics and Public Policy 12,
DOI: 10.1007/978-88-470-5480-6_1, © Springer-Verlag Italia 2014

JEL Classification H42 · I18 · I21 · L13

1 Introduction

Policymakers are increasingly designing policies that encourage patient choice and therefore mobility across providers (hospitals and in some case general practitioners (GPs)). For example, in the English National Health Service (NHS) greater information on measures of hospital quality such as mortality rates, readmission rates, and infection rates are now available on government sponsored websites and patients now have the right to choose any NHS hospital rather than being restricted to hospitals contracting with their local health authority.

When prices are regulated as they are in most countries, health care providers need to compete on quality to attract patients. Patient choice is a pre-requisite for pro-competition policies to improve quality. If patients can freely move among providers then hospitals may have an incentive to compete on quality. The degree to which demand responds to variations in quality plays therefore a critical role. To facilitate choice, policy makers increasingly publish comparative information on health care providers, including measures of the quality such as adjusted mortality and readmission rates. However, patients' choices are also influenced by family doctors and specialists. Family doctors often act as gatekeepers, referring patients to hospitals and specialists who are responsible for diagnosis and treatment. Given the asymmetry of information between doctors and patients, it is not obvious that stimulating choice will have any effect on actual demand.

This chapter reviews the theoretical and empirical literature on patient choice and quality competition in health markets. The theoretical literature summarized in Sect. 2 focuses on key factors that drive incentives to provide quality and how they relate to choice and competition. These include: altruistic motives, cost structure, number of providers, demand responsiveness, GP gatekeeping, degree of specialization, profit constraints and soft budgets. We also review the limited theoretical literature on patients' choice across providers in different countries (e.g. within the EU) or municipalities within the same countries. In Sect. 3 we review selected empirical studies that investigate whether demand responds to quality and waiting times (a negative form of quality), and whether competition affects quality. The latter can be seen as an indirect test of whether providers respond to changes in the degree of patient choice.

2 Theory

The effects of patient choice and competition on health care provider behavior are the subject of an extensive theoretical literature.[1] Here, we focus on the literature that studies this relationship in health care systems where prices are regulated and thus not determined in the market. We therefore do not discuss the literature on patient choice and competition under variable prices, where each provider can set prices constrained only by the demand function it faces or where prices are the result of a bargaining procedure between the purchaser of health services (a private or a public insurer) and the hospital (Barros and Martinez-Giralt 2012).

Most countries regulate prices for health care services. For example, in the English National Health Service under the Payment by Results (PbR) system hospitals are paid a price for each patient which is based on the patient's diagnoses and treatments. The system is similar to the Diagnosis Related Groups system introduced for payments for Medicare patients in the US in the early eighties, variants of which have been adopted subsequently.

It is often claimed that under a fixed-price regime, more patient choice and thus competition lead to higher quality. The intuition is that with fixed prices hospitals can attract more patients only by raising their quality. With more patient choice among hospitals, demand will be more responsive to quality, thereby increasing the additional revenue from raising quality. Formal economic models show that this intuition is correct if (i) more patient choice does increase the responsiveness of demand to quality, (ii) providers are profit maximizer, (iii) the marginal cost of additional patients is constant, (iv) the profit margin (price minus marginal cost) is positive, and (v) providers meet whatever demand is generated by their choice of quality (Ma and Burgess 1993; Gaynor 2006; Gaynor and Vogt 2003).

Thus suppose that the hospital chooses quality q to maximize profit

$$\pi(q;\theta) = T + pX(q;\theta)p - C(X(q;\theta),q) \tag{1}$$

where p is the fixed price, $X(q;\theta)$ is demand (which equals its output) and higher quality increases demand ($X_q > 0$), and C is cost and depends both on output and quality ($C_X > 0$, $C_q > 0$, $C_{qq} > 0$). θ is parameter which measures the degree of patient choice and therefore the responsiveness of demand with respect to quality: $X_{q\theta} > 0$. T is a lump-sum payment from the third party payer. We assume for simplicity that patients' co-payments are set to zero. The choice of quality satisfies the first order condition

$$\pi_q(q;\theta) = [p - C_X(X(q;\theta),q)]X_q(q;\theta) - C_q(X(q;\theta),q) = 0. \tag{2}$$

Quality is increased by patient choice if it increases the marginal profit from quality i.e. if

[1] See Gaynor (2006) and Gaynor and Town (2011) for reviews of the literature on hospital competition.

$$\pi_{q\theta} = (p - C_X)X_{q\theta} - \left(C_{XX}X_q + C_{qX}\right)X_\theta > 0. \tag{3}$$

Patient choice makes demand more responsive to quality ($X_{q\theta} > 0$), but it could increase or reduce the demand for a particular hospital ($X_\theta \lesseqgtr 0$), depending on whether the hospital had higher quality than its competitors, the geographical distribution of patients, and the details of the patient choice policy. Thus a sufficient condition for patient choice policy to increase quality for all providers is that the profit margin ($p - C_X$) is positive, and that the marginal cost of treating an extra patient is not affected by quality, either directly ($C_{Xq} \equiv C_{qX} = 0$) or indirectly ($C_{XX}X_q = 0$).

The incentive to increase quality is stronger, the larger the profit margin. Increasing the regulated price will increase the marginal net profit from higher quality and so will increase quality: the quality supply function is increasing in the regulated price.

In most prospective payment schemes, such as the DRG-pricing schemes and the English PbR scheme, the regulated price is related to the average cost. This implies that the profit margin (price minus marginal cost) will be larger for procedures characterized by large fixed costs and low marginal costs. The profit margin will be positive for hospitals operating at volumes where marginal cost is constant or decreasing. The profit margin will also be greater if the prospective price computation includes investment/capital costs (whether this is the case varies across countries). In some countries, like Norway, the fixed price is a proportion (40–60 %) of the average cost. In this case it is not obvious that the profit margin is positive. If the profit margin is negative, then the financial incentive to increase quality will be negative: providers will wish to reduce quality but will be constrained by sanctions from quality regulators and the threat of malpractice suits from patients. Increases in competition in this case will not increase quality.

2.1 Health care Provider Objectives

The extent to which health care providers respond to competition and patient choice depends on who is taking decisions which affect quality, their preferences, and who is the "residual claimant" (i.e., who has control over any financial surplus). In the English NHS almost all providers offering care to NHS patients are public: they have no shareholders. However, they are subject to financial targets which require them to break even or earn a financial surplus to be reinvested in providing services. Thus they will take the financial consequence of decisions about quality into account.

In most health care providers the decision making process is complex. In hospitals some decisions affecting quality are made at hospital level: hospitals invest in better machinery, medical equipment, or better trained staff. Other decisions, particularly on treatment of patients, are made by individual doctors and nurses who, in many systems of the NHS type, are typically paid a fixed salary, as opposed to fee for service, or a share of profits. We may expect the financial incentive to respond to increased competition by raising quality will be diluted if those who take decisions affecting quality do not receive any financial benefit from higher quality.

In the health economics literature, it is recognized that doctors may not be entirely selfish (McGuire 2000): they care about their income and effort, but they are also motivated in part by an altruistic concern for patients' health. Alternatively, the hospital's objectives can be viewed as the result of bargaining between managers (who are more concerned with financial constraints and targets) and doctors who are more directly concerned with patient wellbeing. It is therefore common in the health economics literature to assume that providers act as if they were maximizing a weighted sum of profits and benefits for the patients.

Brekke et al. (2011) show that with semi-altruistic providers there is an a priori ambiguous relationship between increased patient choice and quality provision. A simplified version of their model illustrates the main point.

The total benefit to the patients who are treated by the provider is $B(X(q;\theta),q)$, with $B_X > 0$ and $B_q > 0$. Higher quality increases patient benefit in two ways: it increases the benefit for each patient treated and it increases the number of patients treated. Suppose that the provider chooses quality to maximize a weighted sum of profit and patient benefit

$$u(q;\theta) = \pi(q;\theta) + \alpha B(X(q;\theta),q) \tag{4}$$

where $\alpha > 0$ is a parameter measuring the degree of provider altruism. Quality is chosen so that

$$u_q = \pi_q(q;\theta) + \alpha\left[B_q(X(q;\theta),q) + B_X(X(q;\theta),q)X_q(q;\theta)\right] = 0. \tag{5}$$

The square bracketed term in (5) is total benefit from higher quality and is positive, so that an altruistic provider will choose a higher level of quality than a pure profit maximizer. Indeed the altruistic provider chooses quality such that $\pi_q(q;\theta) < 0$ which could result in the provider operating where the profit margin $(p - C_X)$ is negative: it treats patients on whom it makes a financial loss.

The effect of increased patient choice on the marginal utility of quality is

$$u_{q\theta} = \pi_{q\theta}(q;\theta) + \alpha\left[\left(B_{qX} + B_{XX}\right)X_\theta + B_X X_{q\theta}\right]. \tag{6}$$

In general the sign of $u_{q\theta}$ is unclear and the effect of patient choice on a profit maximizing provider does not indicate its effect on an altruistic provider. If the marginal costs of output and quality are constant with respect to output and $B_{qX} = 0 = B_{XX}$ then we can ignore the effect of patient choice on demand (X_θ) and

$$u_{q\theta} = \pi_{q\theta}(q;\theta) + \alpha B_X X_{q\theta} = [(p - C_X) + \alpha B_X]X_{q\theta}\,. \tag{7}$$

In this case, increased patient choice has two potentially offsetting effects on the incentives for quality provision. On the one hand, a more quality-responsive demand increases each provider's incentive to reduce quality if $p < C_X$ in order to induce financially unprofitable patients to choose other providers (first term in (7)). On the other hand, altruism means that the provider wants to increase q to attract more patients and increase patient benefit.

Brekke et al. (2011) show that, depending on the size of the two effects, more competition may increase or reduce quality. A similar type of reasoning applies when hospitals compete on waiting times as opposed to quality, where waiting times can be thought of a negative form of quality (Brekke et al. 2008). More competition can lead to higher waiting times when demand is more responsive. Providers have an incentive to increase waiting times to avoid unprofitable patients.

As an additional special case, suppose that the provider cares only about the average quality so that $B_X = 0$. Then, it is still the case that for sufficiently high altruism, providers are willing to choose a quality provision where they make a financial loss by treating an additional patient. In such case, for sufficiently high altruism, a more responsive demand unequivocally leads to lower quality. Providers choose q to deter rather than to attract patients.

2.2 *Profit Constraints*

Hospital care is mainly provided by non-profit and public providers. Thus, large parts of health care services are offered by providers that cannot distribute profits. In publicly-funded health care markets financial surpluses can generally not be distributed in cash to those who makes decisions about provider quality. In such markets it might be more reasonable to model health care providers as being subject to a constraint on profit distribution, where alternative ways of distributing profits (e.g., through perquisites) imply a utility loss relative to distribution in cash. This is the approach taken by Brekke et al. (2012a), who show that the presence of such distribution constraints could severely alter provider incentives with respect to quality. We can incorporate such an approach in the simple example above by assuming that the value of financial profits for the provider is discounted by a factor $\delta \in (0, 1)$.

When investigating the role of profit constraints, it is important to distinguish whether the direct cost of quality are monetary or non-monetary. We therefore now interpret the cost function $C(X(q, \theta), q)$ as the monetary cost of quality and $\varphi(q)$ as the non-monetary cost of providing quality (quality effort disutility), with $\varphi'(q) > 0$ and $\varphi''(q) > 0$.

The condition for optimal quality provision, (5), then becomes

$$(1 - \delta)\left[(p - C_X)X_q(q, \theta) - C_q\right] + \alpha\left[B_q(q, X(q, \theta)) + B_X X_q(q, \theta)\right] = \varphi'(q). \quad (8)$$

It is useful to investigate some special cases. Suppose that there is no altruism and non-monetary costs are also absent ($\alpha = \varphi(q) = 0$). Then the discounting of profits has no effect on quality. The constraint on profit distribution acts like a tax on profits. It reduces both the marginal benefit and the marginal cost of quality, so that the optimal level of quality remains unaffected.

Consider now the case where there is still no altruism but costs are also non-monetary. The provider chooses quality to equate the marginal profit with the marginal non-monetary cost of quality. In this case, profit distribution constraints reduce the marginal profit, and therefore the incentive to provide quality. They also reduce the effect on quality of an increase in the responsiveness of demand to quality (θ).

With semi-altruistic providers, the optimally chosen quality level balances the marginal net profit and the marginal altruistic benefit with the marginal non-monetary cost of quality. A constraint on profit distribution now reduces the importance of the marginal profit from quality ($(1-\theta)\pi_q$) in determining quality. If the marginal effort cost of quality $\varphi'(q)$ is smaller than the altruistic marginal benefit, then $(1-\theta)\pi_q$ is negative in (8) and an increase in δ will reduce the overall marginal loss from higher quality and so increase quality. Conversely if the marginal effort cost of effort is greater than the altruistic marginal benefit, then greater profit constraints reduce quality.

2.3 Soft Budgets

An important feature of many health care systems, is that providers, especially publicly owned hospitals, face soft budgets with funders partially covering deficits or partially confiscating profits. Despite the importance of this topic, there are very few theoretical studies of how soft budgets influence the provision of health care.

Brekke et al. (2013a) analyze the incentives for hospitals to provide quality and expend cost-reducing effort when budgets are soft. They assume that demand is uncertain and that patients can choose which hospital to be treated at. In their model surpluses occur in the low demand state, whereas deficits occur in the high demand state. There are two reasons for this. First, providers cannot increase prices when demand is high (prices are regulated), and, second, they cannot turn down patients who demand treatment. Assuming providers maximize expected profits, the condition for optimal quality is then:

$$\mu(1-\sigma)\left[(p-C_X)X_q^L(q;\theta)-C_q\right]+(1-\mu)(1-\beta)\left[(p-C_X)X_q^H(q;\theta)-C_q\right] = 0, \tag{9}$$

where μ is the probability of the low-demand state (where $X=X^L(q;\theta)$), σ is the share (or probability) of profit expropriation, and β is the share (or probability) of bailout. For a given level of cost-reducing effort, it is evident that profit confiscation reduces the providers' quality incentives and bailout increases quality incentives. However, the effects of soft budgets on quality are ambiguous when providers can expend cost-containment effort (i.e., reduce C_X) to increase their profit margin.

The reason is that softer budgets reduce cost-containment effort, which in turn enhances the negative effect of profit confiscation on quality and counteracts the positive effect of bailouts on quality. Therefore, soft budgets can reduce quality if the effect on cost-containment effort is sufficiently pronounced. Brekke et al. (2013a) show that a higher degree of patient mobility (due to lower travel costs in their Hotelling framework), generally increases quality. A more responsive demand tends on one hand to reduce quality because the profit margin is negative in the high-demand state but on the other hand tends to increase quality because the profit margin is positive in the low-demand state. But the overall expected profit margin in equilibrium is positive and therefore quality increases with a more responsive demand.

2.4 Specialization

As well as competing on quality providers may specialize by attracting particular types of patient. By specializing, providers can reduce the competition they face in their specialist treatment. It is reasonable to assume that specialization is a longer term decision than quality investment. If so, decisions over quality and specialization should be modelled sequentially rather than simultaneously. This is the approach used by Brekke et al. (2006). In stage 1 providers choose the degree of specialization and in stage 2 the level of quality. Suppose that all providers in a market maximize profits ($\alpha = 0$). In stage 2, the optimal quality level, for a given level of specialization s, chosen by each provider is given by (see 5)

$$(p - C_X)X_q(q, s; \theta) - C_q = 0. \tag{10}$$

In a Hotelling spatial competition framework, greater specialization decreases demand responsiveness ($X_{qs} < 0$) and therefore dampens incentives to provide quality ($\partial q/\partial s < 0$). The marginal profit from s when it is chosen at stage 1, is

$$\frac{\partial \pi}{\partial s} = (p - C_X)X_s - C_q \frac{\partial q}{\partial s}. \tag{11}$$

The optimal level of specialization is determined by the marginal cost saving from the effect of greater specialization on quality (second term) and the reduction in profit as market share falls (first term). The degree of specialization critically depends on the degree of convexity of costs with respect to quality. If convexity is high enough, the marginal cost saving due to the reduction in quality as specialization increases is less than the loss in profit as market share falls and so providers do not specialize. Conversely, if the degree of convexity is low, providers specialise to the maximum extent possible.

If there is an interior solution a higher price (p) will increase the marginal benefit of attracting patients and therefore intensify competition and specialization. A higher degree of patient mobility (θ) will make demand more quality-responsive (i.e., increase the value of X_q) and therefore also intensify specialization. The increase in competition intensity following an increase in patient mobility is dampened, but not overturned, by incentives to specialize. Thus, an increase in patient mobility will lead to more specialization and higher quality in equilibrium.

Gravelle (1999) and Nuscheler (2003) investigate quality and provider location decisions (which we can interpret loosely as a type of specialization) when there are no barriers to entry. They find that patient mobility affects entry decisions through incentives for quality provision. Increased patient mobility makes demand more quality-responsive and leads to higher quality provision for a given number of providers. However, this reduces the profit of each provider and therefore reduces entry to the market. Thus, increased patient mobility leads to higher quality but a lower number of providers in equilibrium.

2.5 Information on Quality

The degree of patient mobility, and thus the incentive to compete on quality, may be enhanced by giving more accurate information about quality to patients, for example by publishing data on quality measures. Such information reduces the cost of quality comparisons among hospitals for GPs or patients. It seems intuitive that this would increase the responsiveness of demand to quality and hence increase the marginal profit from increasing quality. However, Gravelle and Sivey (2010) show that this argument is correct only when hospitals do not differ significantly in the quality they provide. If the difference between providers' marginal cost of providing quality is large, then their quality differences will tend to be large. In such cases better information about quality may lead to lower quality. Demand at a hospital depends on the difference in quality between that hospital and a competing hospital, and on the distribution of the differences in the errors each patient makes when observing qualities at the two hospitals. Unless an individual patient's errors in observing quality are perfectly correlated across hospitals, the distribution of the difference in errors for each patient is unimodal and centered on zero. For most patients the difference in their errors is small and few patients have large differences in errors. Thus the marginal revenue from improving quality is smaller, for both hospitals, when the true quality difference is large. Improving the accuracy of patient information makes it even less likely that they will have large differences in errors and thus reduces the marginal revenue from quality increases when the true quality difference is large. Thus if the initial quality difference is large better information will reduce marginal revenue from quality at both hospitals and thus reduce quality levels at both hospitals.

In a related study, Montefiori (2005) assumes that consumers can observe only noisy signals of quality, e.g. true quality with error term, and the error is normally distributed. He shows that patients' utility is decreasing in the variance of the error term and that providers' demand is a function also of the difference in the variance of the error term. He also shows that as long as the variance of the error term is the same across the two providers, the optimal quality incentives for the providers are not affected (since variances drop out from demand functions).

In Gravelle and Masiero (2000) GPs are paid by capitation (a fee for each patient registered in the practice). GPs compete on quality but quality is only imperfectly observed. They show that for a given capitation fee, the presence of imperfect information reduces the effect of a higher capitation fee on quality.

2.6 Gatekeeping

Another way to improve patients' information about secondary care providers is to apply a gatekeeping system, where general practitioners act as gatekeepers to secondary care. Besides the cost containment argument, such a system is likely to ensure that potential patients get access to more accurate information about the characteristics (including quality) of secondary care providers.

Brekke et al. (2007) investigate how hospital quality competition is affected by the introduction of compulsory gatekeeping where every patient needs to have a referral to access a hospital specialist. In their two hospital model patients choose whether to consult a GP who can provide a diagnosis and information about which hospital is a better match for their diagnosis. The first order condition on quality is

$$\lambda\left[(p - C_X)X_q(q;\theta)\right] + (1-\lambda)\left[(p - C_X)\frac{1}{2}\right] - C_q = 0, \tag{12}$$

where λ is the fraction of patients who consult a GP and $1-\lambda$ is the fraction who directly approach the secondary care providers. Patients who consult GPs make an informed choice of providers and therefore this demand segment is responsive to quality (and specialization) differences. Those who do not consult a GP first are uninformed and choose providers randomly with probability 1/2. A gatekeeping system, which corresponds to setting $\lambda = 1$, may amplify or dampen competition depending on the relative size of two effects. On the one hand, competition is amplified by higher GP attendances, i.e., more patients get better recommendations on which provider to use, which makes demand more quality-responsive. On the other hand, incentives for quality competition can be dampened by improved accuracy in diagnosis, which reduces the relative importance of quality as a predictor for the best choice of secondary care provider.

2.7 Cost-Containment Effort

In the absence of cost reimbursement, any type of payment which does not vary with costs (for example a fixed budget or a prospective payment system like PbR) will give the appropriate incentive to keep costs down. If the provider is the residual claimant who has control over any surplus, the benefits from greater cost-reducing effort are equated to the marginal disutility from such effort.

The interaction between quality and cost-containment effort in the presence of patient choice has been analyzed by Brekke et al. (2012a) Assuming that providers are profit maximizers, we can write the optimal conditions for quality and cost-containment effort,

$$(p - C_X)X_q(q;\theta) - C_q = 0, \tag{13}$$

$$-C_e \cdot X(q;\theta) - G_e = 0, \tag{14}$$

where $C_e < 0$ is the reduction in the treatment costs due to cost-containment effort, and G_e is the disutility from exerting such effort. We see that any effect of patient choice on cost-containment effort arises through the interaction with quality. If more patient choice increases quality, and cost-containment effort and quality are substitutes, i.e., $\partial e^*/\partial q < 0$, then more patient choice may lead to lower cost-containment effort. This may be the case when doctor effort can be applied to improving quality of care for individual patients or to finding ways to reduce cost (for example choosing cheaper therapeutically equivalent drugs). Conversely $\partial e^*/\partial q > 0$, if, for example, higher quality leads to higher volume of patients treated, which in turn may increase the marginal benefit from investing in cost-containment effort since the higher profit margin will apply to a larger volume of patients treated (Ma 1994).

2.8 Cream-Skimming

If providers can provide different levels of quality to different patients, then they might have an incentive to cream-skim, i.e., to increase the quality for profitable patients (where the profit margin is positive) and to 'skimp', i.e., to reduce the quality for non-profitable ones (where the profit margin is negative). These incentives may be strengthened in the presence of more intense competition (Ellis 1998). Therefore, if demand is more responsive to quality, providers have even stronger incentives to skimp on expensive patients and to cream-skim the inexpensive ones. Soft budgets will tend to dampen such incentives (Brekke et al. 2013a).

2.9 Interjurisdictional Mobility

A common feature of all the above-discussed analyses is that all health care providers in the market belong to the same health care jurisdiction. In other words, all patient mobility occurs within the same health care jurisdiction. However, in some countries patient mobility across different health care jurisdictions (regions) is an important issue (a prominent example is Italy). Furthermore, patient mobility across national borders is an increasingly important topic in Europe, particularly following recent changes in EU regulations which have strengthened patients' rights to seek medical treatment in other member countries. Brekke et al. (2013b) have examined the implications for increased patient mobility across different health care jurisdictions on quality provision and welfare. They use a spatial competition framework with two regions, where health care quality is decided at regional level (to maximize regional welfare) and patients are allowed freely to choose between the providers in the two regions, which differ in their ability to provide high quality health care.

They find that the effects of patient mobility on quality and welfare in the two regions depend crucially on the details of the transfer pricing scheme; i.e., the price paid by the exporting region (and received by the importing region) for patients who travel across borders to obtain health care. Any increase in quality tends to occur in the region that is the net importer of patients, while quality in the exporting region is reduced for any reasonable assumption regarding transfer pricing. However, with patient mobility across regions a larger share of the total population has access to higher-quality health care, so quality for the *average* patient may increase even if there is no increase in the amount of health care provided by either of the two regions. Allowing for cross-jurisdictional patient mobility also has different welfare effects in the two regions. Intuitively, a higher transfer price tends to favor the importing region at the expense of the exporting region. Importantly, even if allowing for patient mobility might constitute a Pareto improvement at the regional level, where both regions are (at least weakly) better off with mobility than without, such a policy tends always to have winners and losers at the individual patient level.

2.10 Dynamic Analysis

The above analyses use a static framework where providers play a one-shot game, and decisions on quality are made and are implemented in one period. Demand for health care tends to respond sluggishly to changes in quality provision. Because quality is not always easily observable and because of habits or trust in specific health care providers, patients may have sluggish beliefs about quality, which in turn will make demand adjustment sluggish. If a provider increases quality, sluggish beliefs about quality imply that it will take some time before the potential demand increase is fully realized. Therefore, demand may be less responsive when demand is more sluggish.

The implications of demand sluggishness for quality provision are analyzed in a differential-game dynamic setting by Brekke et al. (2012b) and Siciliani et al. (2013). They assume that actual demand vary over time as follows

$$\frac{dD(t)}{dt} = \gamma[D^*(t) - D(t)], \tag{15}$$

where $D(t)$ is actual demand in period t, $D^*(t)$ is potential demand in period t, and $\gamma \in [0, 1]$ is an (inverse) measure of the degree of sluggishness in patients' beliefs about provider quality. The higher is γ, the less sluggish are these beliefs, and the faster is the demand response to quality changes.

In Brekke et al. (2012b) providers are assumed to be profit-maximizers. A lower degree of demand sluggishness will generally stimulate quality provision, regardless of the nature of competition between the providers. A less sluggish demand is analogous to an increase in responsiveness of demand which increases providers' incentives to raise quality. However, the study shows that when providers are allowed to revise their investment decisions more frequently (which can be thought of as a more competitive environment compared to one where quality can be revised only after a long time gap), long-run quality may be lower if the marginal cost of treating a patient is increasing. This result arises because in a dynamic setting lower quality investment by one provider will induce a future reduction in quality investment by the other provider; in other words, quality levels are strategic complements over time. A similar results is derived in Brekke et al. (2010), where demand adjust instantaneously but quality is akin to a stock $q(t)$ which increases over time t only if the investment in quality $I(t)$ is higher than its depreciation rate: $\partial q(t)/\partial t = I(t) - \delta q(t)$. Quality provision is found to be lower in the more competitive environment, where providers are allowed to revise their quality decisions more frequently.

This conclusion contrasts with the findings of Siciliani et al. (2013), who consider the case of partially altruistic health care providers. A reduction in demand sluggishness, for example through public policies that increase the degree of patient information, might lead to lower quality provision if the (exogenous) treatment price is sufficiently low. This result resembles the one found by Brekke et al. (2011) in a static setting (see Sect. 2.1) and the intuition is similar.

3 Empirical Evidence

There are two main empirical approaches to examining the relationship between patient choice and quality of care. The first approach is to test the fundamental assumption that patients choice is influenced by quality and providers with higher quality have a higher demand. Here, quality is an explanatory variable. The second approach tests whether more competition (as measured, for example, by the number of competitors or the Herfindhal index) affects quality. More competition should imply that the demand is more responsive to quality, and therefore competing providers have a stronger incentive to raise quality. Here, quality is the dependent variable.

3.1 Patient Choice and Quality

One of the key steps in the argument that greater competition, with fixed prices, can lead to higher quality is that the choice of a hospital by patients (or their GP advisors) is influenced by its quality relative to that of other available hospitals. Thus by improving quality, relative to other hospitals, a hospital can attract more patients and thereby increase its revenue. The papers are summarized in Table 1. We do not limit the papers to those for markets where price is regulated since we are interested in whether quality affects demand, holding other factors constant, rather than whether competition affects quality.

The analysis is usually conducted for patients in need of a specific treatment (i.e. coronary bypass, PTCA, kidney transplant, cataract surgery, hip replacement). Most of the studies suggest that, after controlling for other factors such as patient characteristics and distance, the probability of a hospital being chosen increases with measures of quality, though the demand elasticities with respect to quality are generally small.

Three of the 12 studies are set in the English NHS (Beckert et al. 2012; Gaynor et al. 2011; Sivey 2008) and are in line with the other, mainly US, studies (Folland 1983; Luft et al. 1990; Burns and Wholey 1992; Hodgkin 1996; Tay 2003; Ho 2006; Howard 2005; Pope 2009). Allowing for waiting times, distance, and patient characteristics, Beckert et al. (2012) and Gaynor et al. (2011) find that higher quality leads to increased demand and Sivey (2008) finds demand increases with quality though not in all his specifications.

Most of the papers model the choice of individuals among hospitals using either conditional logit or mixed logit. The conditional logit model can be motivated within a random utility framework (McFadden 1974). Assume that the utility of patient j choosing hospital i is $U_{ji} = \beta d_{ji} + \gamma q_i + \varepsilon_{ji}$ where d_{ji} is the distance between patient's j residential address and hospital i address, q_i is the quality of hospital j (e.g. mortality rates, readmission rates, waiting times) and ε_{ji} is the unobserved component of utility. If ε_{ji} are iid and follow type 1 extreme value distribution, then the probability of patient j choosing hospital i out of a total of N hospitals is given by:

$$p_{ij} = \frac{e^{(\beta d_{ji} + \gamma q_i)}}{\sum_{i=1}^{N} e^{(\beta d_{ji} + \gamma q_i)}}, \qquad i = 1, \ldots, N. \tag{16}$$

Table 1 Quality and choice of hospital: empirical papers

Paper	Sample	Quality measure	Methods	Covariates	Results	Comments
Beckert et al. (2012)	England 2008/2009. 39,060 elective hip replacement patients, 216 NHS hospitals	Standardized overall mortality rate; MRSA; CQC; good communication	Conditional logit. Similar results from mixed logit with random coefficients	Age, gender, rurality, income, health deprivation. LSOA to site distance; teaching hospital, Foundation Trust, staff per bed, MRSA, CQC quality, CQC financial rating, waiting time; GP relative referral frequency to hospital 2006–2008. PCT effects	Demand decreases with mortality, waiting time, MRSA; increases with CQC rating;	
Burns and Wholey (1992)	Phoenix 1989. 4 medical, 2 surgical DRGs	Standardized in-hospital mortality for each DRG	Conditional logit	Distance. Supply of physicians near hospital	Demand lower if higher mortality for AMI, atrial fibrillation, gastro-intestinal bleeding, large bowel resection	
Cutler et al. (2004)	New York 1991–1999. CABG patients	Reports of standardized mortality	Linear model with time, hospital FEs		Fewer low severity CABG patients at hospital if previous high mortality report	

(continued)

Table 1 (continued)

Paper	Sample	Quality measure	Methods	Covariates	Results	Comments
Gaynor et al. (2011)	England 29 hospitals, 13,000 elective CABG pa, 2003/2004–2007/2008	CABG hospital mortality rate	Conditional logit; waiting time for other procedures as IV; distance as IV for quality	Age, gender, income deprivation, comorbidity; waiting times;	Demand decreases with mortality (0.3 elasticity post 2005); and waiting time (0.3 elasticity post 2005). Sicker patients more sensitive to mortality, less to quality	
Ho (2006)	USA. 1997/1998; 217 hospitals, 11 hospital markets, 28,666 indemnity/PPO patients	Hospital characteristics (teaching status, set of services offered)	Conditional logit; hospital fixed effects; regression of hospital FEs on hospital characteristic	Age, gender, working status, Zip code median income, distance; diagnosis. Hospital dummy characteristics in second stage	Patients prefer hospitals with better services for their diagnosis	No copayment data; 5 % emergency patients
Hodgkin (1996)	27 hospitals. New Hampshire. 1985–1991. Diagnostic cardiac catheterization	Availability of catheterization. Standardized mortality. Staff per bed	Conditional logit. Hospital FE	Age, gender, travel time, comorbidity, primary diagnoses. Patient mortality risk, probability of needing catheterization. Insurance type	Demand increased by catheterization availability for patients most likely to need one. Demand higher if hospital mortality higher	Number of hospitals offering procedure increased from 2 to 9
Howard (2005)	US. Kidney transplant patients 2000–2002	Graft failure rate adjusted for patient characteristics	Mixed logit, random coefficients	Age, gender, ethnicity, diabetes, education, employment, current treatment, insurance type	Demand increases with quality, decreases with distance	

(continued)

Table 1 (continued)

Paper	Sample	Quality measure	Methods	Covariates	Results	Comments
Luft et al. (1990)	California 1983. 7 surgical, 5 medical diagnoses. 99,465 patients. 115 hospitals	Standardized mortality; complication rates	Conditional logit	Distance, charges, medical school, ownership	If mortality higher, demand smaller for 4/7 surgical, 2/5 medical procedures; higher for 1/7 surgical, 1/5 medical	Demand reduced by higher mortality for CABG, AMI
Pope (2009)	USA. 1994–2004. Non emergency Medicare patients	Reported quality rankings	Hospital level demand with year, hospital,speciality FEs; individual level mixed logit (random coefficients)	Distance	Volume increases following improved ranking; probability of choice increases with ranking	
Sivey (2008)	England 2001/2002–2003/2004. 41,019 CABG patients. 33 hospitals	Standardized mortality	One stage model: conditional and mixed logit with standardized mortality. Two stage model: conditional and mixed logit with hospital FEs plus regression of FEs on hospital characteristics including quality	Patient characteristics, diagnoses. Waiting time. Travel time. London hospital	One stage model: demand lower if mortality higher. Two stage model: no significant effect of mortality	

(continued)

Table 1 (continued)

Paper	Sample	Quality measure	Methods	Covariates	Results	Comments
Tay (2003)	AMI. Non HMO Medicare. 14,374 patients. 339 hospitals. California, Oregon, Washington. 1994	10 year average mortality and complication rates; nurses per bed, teaching status, high tech services (catheterization, revascularization)	Conditional logit; mixed logit with random coefficients. Quadratic in mortality, complication rates	Age, gender, ethnicity; beds, distance	Demand falls with distance and increases with quality. But for complication rate demand is higher at low and high rates	50 % patients arrive in ambulance. 46 % patients admitted to hospital nearest to home
Varkevisser et al. (2010)	Netherlands 2003. 5,389 non-emergency first outpatient appointments x neurosurgery. 66 hospitals	Overall reputation; reputation for neurosurgery; University Medical Centre	Conditional logit	Gender, non-adult; self employed. travel time, waiting time below average	Demand increases with overall reputation, waiting time below average. No effect of neurosurgery reputation	Average of 26 hospitals within 60 min
Varkevisser et al. (2012)	Netherlands 2006. 2,670 non emergency Angioplasty (PCI). 19 hospitals. Single health insurer	2005 heart failure readmission rate; pressure sores; overall reputation; cardiology reputation (newspaper ranking)	Conditional logit; mixed logit random coefficients	Age, gender, employment status, travel time. University Medical Centre	Demand increases with overall reputation, cardiology reputation, University Medical Centre; pressure sores; decreases with readmission rate. One point increase in overall reputation (0.4 SD) increases demand by 65 %	Pressure ulcers negatively correlated with other quality measures. No waiting time control

Conditional logit models make the strong assumption that the probability of choosing hospital A rather than hospital B (p_{jA}/p_{jB}) is not affected by the characteristics of other hospitals. Thus the relative probabilities of choosing A over B when the choice set includes hospital C which has the same characteristics as B are assumed to be the same as when the choice set does not include C. But it seems more plausible that patients who would have chosen C will switch to B in its absence so that the proportion choosing A will fall.

Mixed logit models do not impose this restriction and also allow for the possibility that the effects of hospital characteristics on choice vary with unobservable characteristics of patients. The conditional logit model allows the effects of hospital characteristics on choice to vary only with observable characteristics of patients. The mixed logit model is computationally more demanding than the conditional logit model. Tay (2003) shows that the conditional logit underestimates the effect of distance on hospitals' choice.

One of the problems in estimating the effects of hospital quality on choice is that measured hospital quality may depend on the mix of patients at the hospital and the mix of patients may depend in part on quality. Thus, for instance, measured quality is lower if the hospital attracts sicker patients and sicker patients care more about true quality, then the effect of quality on demand will be underestimated. Thus it is important to use measures of quality which allow for the effects of patient characteristics on quality. The papers do this in few ways. First, they may use a standardized quality measure for a hospital calculated as the ratio of actual quality to the quality which would be expected at the hospital given its patients mix. Second, individual patient quality can be regressed on patient characteristics and a hospital dummy and the estimated hospital effect is used as the measure of quality. Both methods are improved by using richer data on individual patients to estimate hospital quality. Third, as in Sivey (2012), demand is first estimated as a function of waiting time, distance and a hospital dummy variable capturing all unobserved hospital factors, and then the hospital effect regressed on hospital characteristics including quality. It is argued (Murdoch, 2006) that this procedure prevents the under-estimation of standard errors which arises if quality measures are entered directly in the demand model. Fourth, to control for time-invariant unobserved heterogeneity (such as the case mix) some studies use panel data including hospital fixed-effects (for example Hodgkin 1996; Tay 2003; Sivey 2012). The approach relies on variations in quality over time to identify an effect on demand. Sivey (2012) for example finds that while cross-sectional results suggest, that higher waiting times lead to an increased probability of hospital choice, panel data results suggest the opposite with higher waiting time reducing the probability of the choice. The counterintuitive cross-sectional results may be explained by unobserved heterogeneity, with hospitals with higher clinical quality attracting higher demand and therefore being characterised by longer waiting times and a higher probability of choice.

A second problem arises in systems, like the NHS, where hospital waiting times adjust to equate demand and supply. Higher treatment quality will lead to longer waiting times if the hospital does not increase the number of patients treated.

Thus waiting times should be included in the demand model because they affect demand and may be correlated with quality. Gaynor et al. (2011) also allow for the possibility that waiting time for coronary bypass (CABG) is correlated with unobserved aspect of CABG quality by using waiting times for other procedures in the same hospital as an instrumental variable for CAGB waiting time. Sivey (2008) estimates a model of waiting time for individual patients including patient characteristics and a hospital dummy and then calculates the waiting time for a hospital as the median of the estimated waiting times. This procedure removes any bias arising from the individual patient quality and waiting times being correlated with unobserved patient characteristics. Beckert et al. (2012) use a hospital level average waiting time which assumes that there are no unobserved hospital level quality measures affecting demand which are also correlated with waiting times.

One way for policy makers to influence the demand responsiveness to quality is to publish information on relative performance, such as risk-adjusted mortality and readmission rates for specific conditions such as CABG. There is evidence from the US that report cards can influence choice and hospitals' market shares in the expected direction. The studies often use a natural-experiment set up since reports cards were introduced in some US states but not in others. For example, Cutler et al. (2004) show that hospitals with high mortality rates experience a 10 % reduction in coronary bypasses but this is not the case for hospitals with low mortality rates.

Mukamel et al. (2005) also finds that higher mortality rates reduce surgeons' market shares. Dranove and Sfekas (2008) find that hospitals with bad reports have a smaller market share but only after accounting for the prior beliefs of the patients: i.e. the market share changes (upwards or downwards) only if the report cards convey information which suggests different quality from that which patients previously expected based on reputation and advice from family doctors. However, report cards may also induce providers to select patients and by avoiding more severe patients (at higher risk of mortality) as evidenced by Dranove et al. (2003).

3.2 Hospital Competition

Most studies on the relation between competition and quality are based on US experiences, but in recent years a number have used English data. The evidence from the US is generally mixed. When prices are fixed (as they are under Medicare in the US), Kessler and McClellan (2000) and Kessler and Geppert (2005) find that a positive effect of competition on quality in the US, Gowrinsankaran and Town (2003) find a negative effect, Shen (2003) finds mixed results, and Shortell and Hughes (1988) and Mukamel et al. (2001) find no effect. Recent evidence from

England suggests that competition increases quality when prices are fixed (Cooper et al. 2011; Gaynor et al. 2010; Bloom et al. 2011a).[2]

The basic empirical approach within a cross-sectional framework is:

$$q_i = \alpha + \beta c_i + \gamma x_i + \varepsilon_i, \quad i = 1, \ldots, N \tag{17}$$

where q_i is the quality provided by hospital i, c_i is a measure of competition, and x_i is a vector of control variables. Quality is typically measures by mortality rates for patients admitted with acute myocardial infarction (AMI) (heart attacks) though occasionally other measures are used. The use of AMI mortality for emergency admissions as a quality measure is justified by its correlation with other measures of quality expected to influence demand for elective care. Competition is typically measured as the number of providers within a catchment area (e.g. 30 min travel time, 40 km) or the Herfindahl–Hirschman index (HHI) which is the sum of the providers' squared market shares. Markets with higher HHIs are interpreted as less competitive.

An observed cross-sectional association between competition and hospital quality does not prove that competition affects quality. For example:

(i) if higher quality hospitals disproportionately attract patients who are sicker and if the quality measure fails to allow appropriately for differences in casemix, and the competition measure is based on market shares, then the observed association between measured quality and measured competition will overestimate the effect of competition on true quality;
(ii) if hospital competition is measured using hospitals' shares of patients treated in an area and patient choice of hospital is influenced by quality, then higher quality hospitals will have greater market shares and will appear to be in less competitive markets, thereby underestimating the effect of competition;
(iii) when competition is measured in terms of the numbers of alternative providers, there may be factors (such as population density or size) which influence both the entry/exit decisions of providers and the quality decisions of providers in the market.

A number of strategies are available to increase the likelihood that an observed association is evidence of causality.

(i) Rich set of covariates. Including covariates which plausibly affect quality reduces the risk of omitted variable bias. In the studies of competition and quality this suggests in particular the need to allow for patient characteristics affecting measured quality i.e. for thorough casemix adjustment. Casemix

[2] See Gaynor (2006), Gaynor and Town (2011), and Propper (2012), for a detailed review of the literature. When prices are variable and endogenously determined, the evidence from the US finds a positive relationship between competition and quality in Sari (2002), Gowrisankaran and Town (2003), and Abraham et al. (2007). However, in England Propper et al. (2004) and Burgess et al. (2008) find a significant (though small) negative relationship for a period in which hospital prices were not centrally regulated.

adjustment by including age, gender etc. as explanatory variables is more flexible than first directly or indirectly standardizing quality with respect to age, gender etc.

(ii) Predicted market shares. Some competition measures are based on patterns of use of providers, as in the Herfindahl–Hirschman index. But observed patterns of use reflect the quality of providers since quality affects patient choice. Thus a provider with higher quality will have larger market share and higher HHI and the estimated effect of competition (smaller HHI) on quality will be under-estimated. Kessler and McClellan (2000) therefore suggest that the HHI is calculated on predicted market shares that arise from estimates based on a demand model of the type described in Sect. 3.1 which includes distance as the key explanatory and which does not contain hospital quality or other hospital variables correlated with quality.

(iii) Instrumental variables for competition: variables which are correlated with competition but which are not correlated with unobservable factors affecting quality. For example, Bloom et al. (2011a) use the political marginality of an area as an instrument for the competing number of hospitals, arguing that decisions on closing or merging hospitals will be affected by political considerations (public hospitals are less likely to be closed in marginal constituencies) is used which are independent of the quality a hospital. In their study on the effect of competition on management quality they have 2005/2006 cross-sectional data for 100 trusts (about 61 % of all NHS trusts). The key measure of competition is the number of hospitals within a 30 km radius. They find that increasing the number of rival providers by three increases the index of management quality by more than a standard deviation, which implies a 6 % reduction in AMI mortality rates.

(iv) Difference in differences: compare changes in quality over time for providers which had different changes in competition. The obvious way to implement this strategy is compare hospitals where rivals entered or left with those where the set of rivals was unchanged. Alternatively, one can argue that competition depends both on market structure (numbers of rivals or market shares) and on national competition policy and that changes in national policy lead to different changes in effective competitive pressures for providers depending on their market structure. The methodology of difference-in-difference relies on the assumption of common trends: quality would have changed at the same rate for providers with different number of competitors in the absence of the change in structure or competition policy. If the trends differ, then the estimated effect of greater competition will be biased upwards if quality improves more rapidly in more competitive areas. It is possible, as in Cooper et al. (2011), to allow differential trends in the high and low competition groups of hospital and to test if the change in policy affected the trend one group compared to the other.

Cooper et al. (2011) estimates the effect of competition on AMI mortality rates during a period (2002–2008) when prices were fixed. The analysis uses data from

227 hospital sites as opposed to data aggregated at Trust level (therefore allowing for Trusts with multiple sites). They assume that the introduction of the Patient Choice policy from 2005 onward led to an increase in competition and that this led to an increase in effective competition which was greater in areas with more providers. Competition was measured using market shares based on predicted rather than actual demand. They estimate a difference-in-difference model and find that the introduction of Patient Choice policy was associated with a bigger reduction in AMI mortality in more competitive areas. AMI mortality fell by 0.31 % points per year faster in areas where competition was more intense by one standard deviation of their market structure indicator.

Gaynor et al. (2011) use a similar methodology, also arguing that the introduction of Patient Choice increased competition more for hospitals in markets with a more competitive structure and using predicted demand to calculate market shares. They use two years of data: 2003 and 2007. The sample includes 130 hospitals (trusts) in 2003 and 121 in 2007. They find that hospitals which had a 10 % higher Herfindahl index (i.e. faced less competition) in 2003 had a smaller (by 2.9 %) decrease in AMI mortality rates between 2003 and 2007. There is a similar association for overall mortality rates, though the effect is quantitatively smaller. Hospitals facing less competition in 2003 also had an increase in overall length of stay between 2003 and 2007 relative to providers facing more competition in 2003.[3]

4 Conclusions

Patient choice and mobility across providers (such as hospitals and GPs) are likely to become permanent features of health systems in several countries. This chapter has focussed on the theoretical and empirical literature on patient choice and quality competition when prices are regulated. The theoretical literature suggests that more choice increases quality if providers are profit maximizers and the marginal cost is constant. The results are less clear-cut when providers are altruistic and the marginal cost of treatment is increasing. For sufficiently high altruism, providers may be willing to work at a negative profit margin and more choice may then reduce quality. We have also outlined that profit constraints may dampen incentives to compete on quality but, if providers are altruistic, may contribute to increase the relative weight given to patients so that the overall effect on quality is indeterminate. Hospitals may also dampen incentives to compete on quality by specializing. When considering patients mobility across different regions (or countries) transfer pricing schemes between the importing and exporting regions

[3] Bevan and Skellern (2011) and OHE (2012) have also reviewed these studies and the three most recent ones have been subject of an exchange in The Lancet (Pollock et al. 2011a, b; Bloom et al. 2011b, 2012).

play a critical role to align regions' incentives. The empirical literature generally confirms that demand responds to quality and waiting times though demand elasticities are generally low. Recent empirical studies mainly from the US and the UK generally confirm that more competition among hospitals is positively related to quality.

References

Abraham, J. M., Gaynor, M., & Vogt, W. B. (2007). Entry and competition in local hospital markets. *Journal of Industrial Economics, 2*, 265–288.

Barros, P. P., & Martinez-Giralt, X. (2012). Models of negotiation and bargaining in health care. In A. Jones (Ed.), *The Elgar companion to health economics* (pp. 231–239). Cheltenham: Edward Elgar.

Beckert, W., Christensen, M., & Collyer, K. (2012). Choice of NHS-funded hospital services in England. *Economic Journal*. doi: 10.1111/j.1468-0297.2012.02496x.

Bevan, G., & Skellern, M. (2011). Does competition between hospitals improve clinical quality? A review of evidence from two eras of competition in the English NHS. *British Medical Journal, 343*, d6470.

Bloom, N., Propper C., Seiler S., & Van Reenen, J. (2011a). *The impact of competition on management quality: Evidence from public hospitals*, CEP Discussion Paper N. 983.

Bloom, N., Cooper, Z., Gaynor, M., Gibbons, S., Jones, S., McGuire, A., et al. (2011b). In defence of our research on competition in England's national health service. *The Lancet*, December 17/24/31:2064–2065.

Bloom, N., Cooper, Z., Gaynor, M., Gibbons, S., Jones, S., McGuire, A., et al. (2012). A response to Pollock et al (2011). http://www.bristol.ac.uk/cmpo/publications/other/cpresponse.pdf.

Brekke, K. R., Nuscheler, R., & Straume, O. R. (2006). Quality and location choices under price regulation. *Journal of Economics and Management Strategy, 15*, 207–227.

Brekke, K. R., Nuscheler, R., & Straume, O. R. (2007). Gatekeeping in health care. *Journal of Health Economics, 26*, 149–170.

Brekke, K. R., Siciliani, L., & Straume, O. R. (2008). Competition and waiting times in hospital markets. *Journal of Public Economics, 92*, 1607–1628.

Brekke, K. R., Siciliani, L., & Straume, O. R. (2011). Hospital competition and quality with regulated prices. *Scandinavian Journal of Economics, 113*, 444–469.

Brekke, K. R., Siciliani, L., & Straume, O. R. (2012a). Quality competition with profit constraints. *Journal of Economic Behavior and Organization, 84*, 642–659.

Brekke, K. R., Siciliani, L., & Straume, O. R. (2013a). *Hospital competition with soft budgets*. Mimeo, Braga, Portugal.

Brekke, K. R., Cellini, R., Siciliani, L., & Straume, O. R. (2010). Competition and quality in regulated markets: A differential-game approach. *Journal of Health Economics, 29*, 508–523.

Brekke, K. R., Cellini, R., Siciliani, L., & Straume, O. R. (2012b). Competition in regulated markets with sluggish beliefs about quality. *Journal of Economics and Management Strategy, 21*, 131–178.

Brekke, K. R., Levaggi, R., Siciliani, L., & Straume, O. R. (2013b). *Patient mobility, health care quality and welfare*. Mimeo, Bergen, Norway.

Burgess, S., Propper, C., & Gossage, D. (2008). Competition and quality: Evidence from the NHS internal market 1991–9. *The Economic Journal, 118*, 138–170.

Burns, L. R., & Wholey, D. R. (1992). The impact of physician characteristics in conditional choice models for hospital care. *Journal of Health Economics, 11*, 43–62.

Cooper, Z., Gibbons, S., Jones, S., & McGuire, A. (2011). Does hospital competition save lives? Evidence from the English patient choice reforms. *Economic Journal, 121*(554), F228–F260.

Cutler, D., Huckman, R., & Landrum, M. B. (2004). The role of information in medical markets: An analysis of publicly reported outcomes in cardiac surgery. *American Economic Review, 94*, 342–346.
Dranove, D., Kessler, D., McLellan, M., & Satterthwaite, M. (2003). Is more information better? The effects of 'report cards' on health care providers. *Journal of Political Economy, 111*(3), 555–588.
Dranove, D., & Sfekas, A. (2008). Start spreading the news: A structural estimate of the effects of New York hospital report cards. *Journal of Health Economics, 27*(5), 1201–1207.
Ellis, R. (1998). Creaming, skimping and dumping: Provider competition on the intensive and extensive margins. *Journal of Health Economics, 17*(5), 537–555.
Folland, S. T. (1983). Predicting hospital market shares. *Inquiry, 20*, 34–44.
Gaynor, M. (2006). What do we know about competition and quality in health care markets? *Foundations and Trends in Microeconomics*, *2*(6), 441–508.
Gaynor, M., & Vogt, W. B. (2003). Competition among hospitals. *RAND Journal of Economics, 34*, 764–785.
Gaynor, M., Moreno-Serra R., & Propper, C. (2011). *Death by market power: Reform, competition and patient outcomes in the national health service*, CMPO Working Paper No. 10/242.
Gaynor, M., Propper, C., & Seiler, S. (2010). *The effect of patient choice*. Evidence from recent NHS reforms. cep.lse.ac.uk/conference_papers/10-05-2010/seiler.pdf.
Gaynor, M., & Town R. (2011). Competition in health care markets. In M. V. Pauly, T. G. McGuire, P. P. Barros (Eds.), *Handbook of health economics* (Vol 2). Elsevier, Amsterdam.
Gowrisankaran, G., & Town, R. (2003). Competition, payers, and hospital quality. *Health Services Research, 38*, 1403–1422.
Gravelle, H. (1999). Capitation contracts: Access and quality. *Journal of Health Economics, 18*, 315–340.
Gravelle, H., & Masiero, G. (2000). Quality incentives in a regulated market with imperfect competition and switching costs: Capitation in general practice. *Journal of Health Economics, 19*, 1067–1088.
Gravelle, H., & Sivey, P. (2010). Imperfect quality information in a quality-competitive hospital market. *Journal of Health Economics, 29*, 524–535.
Hodgkin, D. (1996). Specialized service offerings and patients' choice of hospital: The case of cardiac catheterization. *Journal of Health Economics, 15*, 305–332.
Ho, K. (2006). The welfare effects of restricted hospital choice in the US medical care market. *Journal of Applied Econometrics, 21*, 1039–1079.
Howard, D. H. (2005). Quality and consumer choice in health care: Evidence from kidney transplantation. *Topics in Economic Analysis and Policy, 5*(1), 1349.
Kessler, D., & McClellan, M. (2000). Is hospital competition socially wasteful? *The Quarterly Journal of Economics, 115*, 577–615.
Kessler, D. P., & Geppert, J. J. (2005). The effects of competition on variation in the quality and cost of medical care. *Journal of Economics and Management Strategy, 14*(3), 575–589.
Luft, H. S., Garnick, D. W., Mark, D. H., Peltzman, D. J., Phibbs, C. S., Lichtenberg, E., et al. (1990). Does quality influence choice of hospital? *JAMA, 263*, 2899–2906.
Ma, C. A., & Burgess, J. F. (1993). Quality competition, welfare, and regulation. *Journal of Economics, 58*, 153–173.
Ma, C. A. (1994). Health care payment systems: Cost and quality incentives. *Journal of Economics and Management Strategy, 3*, 93–112.
McFadden, D. (1974). Conditional logit analysis of qualitative choice behaviour. In P. Zarembka (Ed.), *Frontiers in econometrics* (4, pp. 105–142). New York: Academic Press.
McGuire, T. (2000). Physician agency, In A. J. Culyer & J. P. Newhouse (Eds.), *Handbook of health economics* (Vol. 1). Amsterdam: Elsevier.
Montefiori, M. (2005). Spatial competition for quality in the market for hospital care. *European Journal of Health Economics, 6*, 131–135.

Mukamel, D., Zwanziger, J., & Tomaszewski, K. J. (2001). HMO penetration, competition and risk-adjusted hospital mortality. *Health Services Research, 36*, 1019–1035.

Mukamel, D. B., Weimer, D. L., Zwanziger, J., Gorthy, S. F., & Mushlin, A. I. (2005). Quality report cards, selection of cardiac surgeons, and racial disparities: A study of the publication of the New York state cardiac surgery reports. *Inquiry, 41*, 435–446.

Murdoch, J. (2006). Handling unobserved site characteristics in random utility models of recreation demand. *Journal of Environmental Economics and Management, 51*,1–25.

Nuscheler, R. (2003). Physician reimbursement, time consistency, and the quality of care. *Journal of Institutional and Theoretical Economics, 159*(2), 302–322.

OHE, (2012). Report of the Office of Health Economics Commission on Competition in the NHS. London: Office of Health Economics.

Pollock, A., Macfarlane, A., Kirkwood, G., Majeed, A., Greener, I., Morrelli, C., et al. (2011a). No evidence that patient choice in the NHS saves lives. *The Lancet, 378*, December 17/24/3, 2057–2060.

Pollock, A., Majeed, A., Macfarlane, A., Greener, I., Kirkwood, G., et al. (2011b). Authors' reply. *The Lancet, 378*, December 17/24/31, 2065–2066.

Pope, D. (2009). Reacting to rankings: Evidence from "America's best hospitals". *Journal of Health Economics, 28*, 1154–1165.

Propper, C. (2012). Competition, incentives and the English NHS. *Health Economics, 21*, 33–40.

Propper, C., Burgess, S., & Green, K. (2004). Does competition between hospitals improve the quality of care? Hospital death rates and the NHS internal market. *Journal of Public Economics, 88*(7–8), 1247–1272.

Sari, N. (2002). Do competition and managed care improve quality? *Health Economics, 11*, 571–584.

Shen, Y.-S. (2003). The effect of financial pressure on the quality of care in hospitals. *Journal of Health Economics, 22*, 243–269.

Shortell, S. M., & Hughes, E. F. (1988). The effects of regulation, competition, and ownership on mortality rates among hospital inpatients. *New England Journal of Medicine, 318*, 1100–1107.

Siciliani, L., Straume, O.R., and Cellini, R. (2013). Quality competition with motivated providers and sluggish demand, *Journal of Economic Dynamics and Control. 37*, 2041–2061, forthcoming.

Sivey, P. (2008). *The effect of hospital quality on choice of hospital for elective heart operations in England.* Chapter 4 in The Economics of Patient Choice and Information in the National Health Service, Ph.D. Thesis. University of York, York.

Sivey, P. (2012). The effect of waiting times and distance of hospital choice for English cataract patients. *Health Economics, 21*, 444–456.

Tay, A. (2003). Assessing competition in hospital care markets: The importance of accounting for quality differentiation. *RAND Journal of Economics, 34*, 786–814.

Varkevisser, M., van der Geest, S., & Schut, F. (2010). Assessing hospital competition when prices don't matter to patients: The use of time elasticities. *International Journal of Health Care Finance and Economics, 9*(4), 347–366.

Varkevisser, M., van der Geest, S., & Schut, F. (2012). Do patients choose hospitals with high quality ratings? Empirical evidence from the market for angioplasty in the Netherlands. *Journal of Health Economics, 31*, 371–378.

Using Discrete Choice Experiments to Understand Preferences in Health Care

Christian Pfarr, Andreas Schmid and Udo Schneider

Abstract Whenever processes are reconfigured or new products are designed the needs and preferences of patients and consumers have to be considered. Although at times neglected, this becomes more and more relevant in health care settings: Which modes of health care delivery will be accepted? What are the patients' priorities and what is the willingness to pay? To which degree are patients mobile and for which kind of services are they willing to travel? Preferences, however, are difficult to measure, as they are latent constructs. This becomes even more difficult, when no past choices can be analyzed either as the service or the product is yet to be developed or as in the past there has not been free choice for patients. In such cases, preferences cannot be surveyed directly. Asking individuals openly for their attitudes towards certain services and products, the results are likely biased as individuals are not confronted with budget constraints and trade-offs. For this reason, discrete choice experiments (DCEs) are frequently used to elicit patient preferences. This approach confronts patients with hypothetical scenarios of which only one can be chosen. Over the past few years, this tool to reveal patients' preferences for health care has become very popular in health economics. This contribution aims at introducing the principles of DCEs, highlighting the underlying theory and giving practical guidance for conducting a discrete choice experiment in health economics. Thereby we focus on three major fields of patient demand: designing health insurance, assessing patient utility of new

C. Pfarr (✉)
University of Bayreuth, Universitätsstrasse 30, 95440 Bayreuth, Germany
e-mail: christian.pfarr@uni-bayreuth.de

A. Schmid
JP Health Management, University of Bayreuth, Universitätsstrasse 30 95440 Bayreuth, Germany
e-mail: andreas.schmid@uni-bayreuth.de

U. Schneider
WINEG—Scientific Institute of TK for Benefit and Efficiency in Health Care, Bramfelder Straße 140 22305 Hamburg, Germany
e-mail: Dr.Udo.Schneider@wineg.de

R. Levaggi and M. Montefiori (eds.), *Health Care Provision and Patient Mobility*, Developments in Health Economics and Public Policy 12,
DOI: 10.1007/978-88-470-5480-6_2,

pharmaceuticals and analyzing provider choice. By having a closer look at selected international studies, we discuss the application of this technique for the analysis of the supply and the demand of health care as well as the implications for assessing patient mobility across different health care systems.

Keywords Measuring preferences · Discrete choice experiment · Demand for health care

JEL Classification I11 · I13 · C93

1 Introduction

Patients' mobility is always the result of a decision making process. In economic theory patients' criteria for making such a decision are the benefit which they gain and the costs that they have to bear. They decide on the basis of their preferences and the scarce resources that they have available. At the very end, they will opt for the alternative which helps them to achieve the highest overall level of utility.

In the case of mobility in health care the benefits of being mobile, i.e. traveling, may include access to a better health care provider or getting quicker treatment. However, aspects such as the quality of a health care provider embraces various factors such as medical skills, friendliness or ancillary services. Even if we could agree on a number of factors and a scale for each of them to measure the quality it would still be difficult to weight these factors against each other. Are medical skills two times or five times as important as the friendliness of the caregiving staff? Are patients at all willing to trade off these factors? The same holds true for the costs of traveling. There are different direct and indirect costs, depending on the mode of traveling, i.e. if a car, public transport or a taxi is chosen. These options also go along with different levels of comfort and different durations of the travel. Again—how much would the mobility of patients be increased if the direct costs of traveling could be reduced by let's say 10 %? How about the opportunity costs?

One way to find out about patients mobility and the relevance of the different factors that are involved in the decision making process would be to ask the patients directly: Rate on a scale from 1 to 10 the importance of the following provider characteristics, i.e. medical skills, friendliness etc. This is a good example for the problems that come with such an approach: Everything is important! And as long as you are not forced to trade off these alternatives the chance is high that patients' wishes are "inflated".

In the following we will introduce discrete choice experiments (DCEs) as a method that allows to overcome these obstacles. Although the method per se has a long history its application in the health care setting is still rather young. For this reason, the first part of this chapter is dedicated to the theoretical and empirical fundamentals of this technique. The second part highlights some of the potentials and limitations of DCEs by discussing some applications in the fields of health insurance and pharmaceuticals before turning to some articles that bear a closer relation to patient mobility, i.e. articles that analyze provider choice.

2 Measuring Preferences

When analyzing individuals' preferences the first decision is about which concept of preferences should be applied. In general, two different approaches exist i.e. *revealed* and *stated preferences. Revealed preferences* rely on the utility concept of Samuelson (1938) which draws on observable actual market behavior. Accordingly, the observed behavior of humans leads to theory consistent estimates of their demand function. Typically, factors such as the price and the quantity of the product are used. As a consequence, *revealed preference* data are highly reliable as the actual purchase of the good documents individuals' preferences (Louviere et al. 2000). However, there are two main reasons for using *stated* rather than *revealed* preferences in health economics, i.e. (Louviere et al. 2000):

- It may be necessary to approximate the demand for a new product that does not yet exist, i.e. is a hospital provider aiming at opening a new hospital at a new location. For his decision he needs information about individuals' preferences: Will they make use of this new facility given for example the distance from their place of residence and their mobility or not?
- The good may not be traded in real markets. Especially in the health care sector economists frequently have to deal with a lack of market prices. For instance, in cases of a mandatory social health insurance the contribution rate is usually exogenous and does not reflect an individual's preferences.

Stated preference data are able to reflect decisions between alternative, hypothetical goods and enable researches to forecast individuals' decision making by revealing existing but not yet articulated preferences (Louviere et al. 2000). In contrast to *revealed preference* data, this concept gathers data by confronting individuals with hypothetical scenarios in an experimental survey. Thus the two concepts can be distinguished by how the data are collected.

Discrete choice experiments (DCE) are one method that can be used to collect stated preference data.[1] Louviere and Woodworth (1983) as well as Louviere and Hensher (1982) have contributed significantly in developing DCEs in their current form. The method of DCEs was first applied in environment and transportation economics and since the middle of the 1990s also in health economics with growing popularity. Whereas only about 34 DCEs were conducted between 1990

[1] Another methodology frequently applied in health economics is a *Contingent Valuation*. For more information about this concept please refer to Mitchell and Carson (1989) or Johannesson (1996).

and 2000, the literature review of De Bekker-Grob et al. (2012) counts about 114 DCEs between 2001 and 2008.

2.1 Theory of Discrete Choice Experiments

Before we start with the theoretical background of DCEs, we give a first quick overview about how these experiments work.[2] In general, products or goods can be specified by their characteristics i.e. *attributes*. In line with the example above, the attributes of a hospital could be distance, number of clinics, quality etc. The attributes also differ with respect to their values i.e. *attribute levels*. Accordingly, the attribute distance could be measured in different levels of miles. By varying the attribute levels one gets new, hypothetical products, the so called *alternatives*. These alternatives are usually graphically visualized on show cards. In this framework, preferences are measured by confronting respondents with two or more alternatives (*choice sets*) and by asking them which one they would choose. Usually, respondents have to choose repeatedly between varying alternatives. In line with the underlying decision theory, rational individuals will always choose the alternative offering the highest utility. Utilizing the thereby generated data the methodology of DCE is able to reveal individuals' preferences.

The theoretical background of DCEs is based on the new demand theory and its characteristics approach of Lancaster (1966). In contrast to the neoclassical theory, Lancaster suggests that individuals do not derive utility from the consumption of goods per se. On the contrary, the new consumer approach assumes individuals to benefit from the attributes of the goods. This means that a patient does not generate utility from the visit of a hospital per se, but from the services a hospital provides i.e. the specific skills of its practitioners or the general level quality or friendliness etc. Although the new consumer approach focuses on attributes rather than whole goods, the theory is conform to neoclassic utility theory regarding the form and shape of the utility function. Therefore, the axioms of transitivity, completeness, continuity and concavity hold (Lancaster 1971). The original consumer approach concentrates on an optimal combination of attributes by choosing a (potentially large) number of different goods. In contrast, DCEs require individuals to choose only one option out of two or more alternatives. Thus, DCEs create a specialized form of the general consumer approach.

This general concept, i.e. that each good is characterized through the relevant attributes as well as the desired attribute levels that affect individuals' utility, can be formalized. A rational individual i maximizes his utility over a combination of attributes z. As good j can consist of various attributes, the combination of the

[2] The following paragraphs draw on Pfarr (2013) and Pfarr and Schmid (2013). For more information on this paragraph see Amaya-Amaya et al. (2008), Bateman et al. (2002) and Pfarr (2013).

attribute vector b_j and the quantity indicator x_j (attribute levels) leads to z_j, i.e. the bundle of attributes of good j (Lancaster 1971).

$$U_i = u_i(z) = u_i(b, x) \qquad \text{with } z_j = b_j x_j; \qquad j = 1, \ldots, n. \tag{1}$$

As attributes vary regarding their attribute levels, new hypothetical alternatives can be created by combining the attributes with different levels. At the end one gets a set of distinct—hypothetical—alternatives which are presented to the respondents within a DCE. The alternatives can be grouped together in a way that respondents have to choose between two, three or more alternatives. In line with the underlying decision theory, utility maximizing individuals will always choose the alternative offering the highest utility.

More technically, an individual i will only choose alternative l if the utility derived from alternative l exceeds the utility derived from any other alternative j (Ben-Akiva and Lerman 1985; Louviere et al. 2010). This leads to:

$$U_{ij} = u_i\left(x_j; b_j\right) \rightarrow \max! \tag{2}$$

Equation (2) states that an individual derives his utility from the attributes of the good j and the quantity of these attributes x_j. From the maximization of this utility function the conditional demand function x_j of alternative j can be derived:

$$x_j(p_j; b_j; y_i; s_i) \tag{3}$$

In combination with Eq. (2) this leads to the indirect (conditional) utility function of individual i, v_i.

$$V_{il} > V_{ij} \Leftrightarrow v_l(p_l, b_l, y_i, s_i) > v_j(p_j, b_j, y_i, s_i) \quad \forall j \neq l. \tag{4}$$

The utility function consists of the price of the respective alternative p_l, which is basically just another attribute of the alternative, the remaining attributes are indicated by b_l, individuals' income by y_i and his socio-demographic characteristics by s_i. From Eq. (4) it is obvious that we do not calculate the utility for each individual. Instead, the theory of a representative consumer implies that only the aggregated preferences of a group of individuals are considered (Anderson et al. 1992). However, reality suggests that individuals differ regarding their preferences. To be able to analyze heterogeneous preferences the vector s_i is incorporated in the utility function and controls for a number of observable (socio-demographic) characteristics.

Obviously it is not possible to observe ex ante all determinants affecting individuals' utility. There will always be factors such as unaccounted attributes, latent variation in utilities between individuals or measurement errors which will be unobservable to the researcher. Accordingly, utility is a deterministic argument for the individual but a stochastic one for the researcher (Anderson et al. 1992; Manski and Lerman 1977) can only observe individuals decisions regarding different alternatives. Thus, utility is a latent construct. Consequently and in line with the underlying random utility theory by McFadden (1974, 1981) and Manski

(1977), an error term ε_{il} is added to the indirect utility function. Hence, the utility function is stochastic and can be additively split into a deterministic observable part $w_l(\bullet)$ and a stochastic component ε_{il}:

$$V_{il} > V_{ij} \Leftrightarrow w_l(\bullet) + \varepsilon_{il} > w_j(\bullet) + \varepsilon_{ij} \qquad \forall j \neq l. \tag{5}$$

Therefore, only the probability P_{il} of individuals choosing alternative l rather than any other alternative j can be estimated (Louviere and Street 2000). The estimated probability indicates individuals' decision making and corresponds to their demand for a given good expressed by choosing one of the proposed options. This leads to:

$$\begin{aligned} P_{il}(l|C_m) &= \Pr\left[\varepsilon_{ij} - \varepsilon_{il} < wl(\bullet) - wj(\bullet)\right] \; \forall 1, \mathrm{j} \in C_m; \forall j \neq l \\ &= \int\limits_{\varphi il} \left[\varepsilon_{ij} - \varepsilon_{il} \leq w_l(\bullet) - w_j(\bullet)\right] \phi(\varphi_{il}) d\varphi_{il}, \end{aligned} \tag{6}$$

with $\varphi_{il} = \varepsilon_{ij} - \varepsilon_{il}$. Therefore the probability is equal to the probability that differences between the error terms are dominated by differences in the deterministic component (Train 2009). As every respondent is forced to decide repeatedly between a predefined set of alternatives, a panel structure can be assumed. Depending on the assumptions made about the error terms one obtains a logit or a probit specification. In the case of a binary discrete decision between two alternatives, Eq. (6) would lead to a binary probit or logit model whereas a choice situation between more than two alternatives is typically modeled as a multinomial logit model.[3] The dependent variable in such models always indicates the individuals' decisions and the independent variables cover the attributes as well as the socio-demographic characteristics.

Figure 1 visualizes the choice process as described in this chapter. Within a DCE, individuals choices are the result of a sequential decision making process while making comparisons between different alternatives and dealing with trade-offs (Amaya-Amaya et al. 2008). Figure 1 shows a simplified binary choice between two attributes m and n.

If a baseline alternative is included—i.e. typically the status quo S—a rational individual will only choose a proposed alternative B if this alternative offers a higher utility than the status quo. If at the same time the individual chooses status quo S rather than alternative A, the individual's indifference curve must be located between A and B (Pfarr 2012; MacNeil Vromen and Zweifel 2011). In the course of the experiment, each respondent has to choose repeatedly between varying alternatives, which allows the approximation of the indifference curve. It is very important that the individuals are forced to "jump back and forth" between the proposed alternatives indicating a higher or lower utility level (Zweifel et al. 2010). As the slope of the indifference curve is $-\Delta m/\Delta n$, the marginal rate of

[3] For detailed information on different estimation models refer to Train (2009), Ben-Akiva (1985) or Cameron and Trivedi (2009).

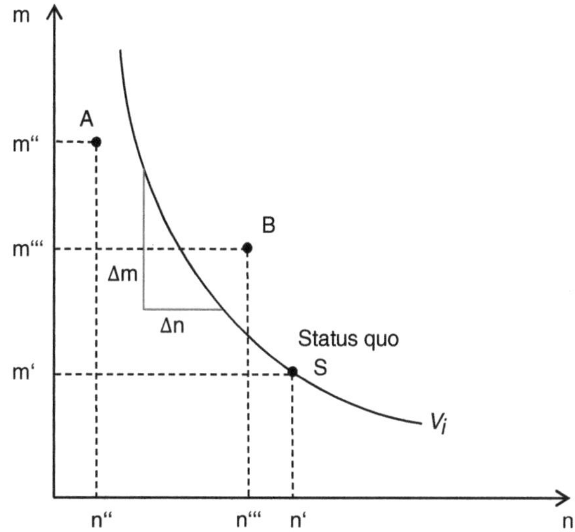

Fig. 1 Choice situation and approximation of the indifference curve. *Source* MacNeil Vroomen and Zweifel (2011), own visualization

substitution $MRS_{n,m}$ between these two attributes can be easily computed. The estimated parameters of the indirect utility function reflect the marginal utilities of the respective attributes and the $MRS_{n,m}$ is given by (Lancsar et al. 2007):

$$MRS_{b_n}^{b_m} = -\frac{\partial v_l(p_l, b_l, y_i, s_i)/\partial b_m}{\partial v_l(p_l, b_l, y_i, s_i)/\partial b_n} = -\frac{\hat{\delta} b_m}{\hat{\delta} b_n} \tag{7}$$

Furthermore, if we substitute ∂b_n by the price attribute ∂p_l the *MRS* can be interpreted as marginal willingness-to-pay (MWTP).[4] This is the MWTP of individual *i* for an additional unit of b_n expressed in units of individuals' income. MWTP values are often calculated in the analysis of DCEs to simplify the interpretation of the results.

2.2 Practical Issues in Conducting a DCE

A DCE measures preferences over attributes in hypothetical decision situations. Accordingly, we need an experimental setup which incorporates the relevant information affecting individuals' utility. An adequate setup can be developed according to the guidelines presented in Bateman et al. (2002). The authors suggest five steps in developing a DCE which we will discuss below in more detail. These are:

---

[4] The price parameter can be interpreted as the marginal utility of income with the help of Roy's Identity. For formal proof see Hanemann (1983) or Telser (2002).

1. identifying attributes;
2. assigning attribute levels;
3. applying an experimental design;
4. generating and visualizing choice sets;
5. collecting data.

2.2.1 Identifying Attributes

The first step is to identify all relevant attributes affecting individuals' decisions and utility respectively. In general, all factors—qualitative as well as quantitative attributes—are to be considered. Telser (2002) suggests that the chosen attributes should have a high relevance for individuals' decisions. This reduces the impact of the stochastic component of the utility function specified above. According to Bateman et al. (2002), depending on the research question, an attribute can gain relevance if it can be influenced by policy or decision makers. Thus, attributes could also be chosen strategically. Finally, attributes should not be mutually exclusive. Instead they should be in a substitutive relationship to each other allowing for the implementation of trade-offs (Telser 2002). For welfare measures such as the marginal willingness-to-pay it is of particular importance to incorporate a price attribute (*payment vehicle*). A *payment vehicle* within a health related choice experiment could probably be insurance contribution, amount of cost sharing or distance to the nearest hospital (Gyrd-Hansen and Skjoldbordg 2008; Johnson et al. 2011).

There is no golden rule on how many attributes are recommended. This always depends on the research question. However, unless very large samples can be surveyed a restriction is necessary to limit the cognitive burden respondents have to deal with when making their decisions (Bateman et al. 2002). A starting point should be an intensive literature review of already existing experiments complemented by expert interviews and focus group analysis to uncover the attributes that are of top priority. Kløjgaard et al. (2012) provide interesting insights into the relevance of qualitative research prior to a quantitative pre-test.

At the end there are only rules of thumb of how many attributes should be chosen. According to Guttmann et al. (2009) and De Bekker-Grob et al. (2012), on average five to six attributes were chosen in experiments conducted between 2001 and 2008.

2.2.2 Assigning Attribute Levels

In a second step, the levels of the attributes have to be defined. Basically there is no need to assign every attribute the same number of levels. With two levels, only linear marginal utility can be estimated, while a higher number of levels allows the researcher to model more complex non-linear utility functions (Amaya-Amaya et al. 2008). In general, the more levels are assigned the higher is the chance that an accurate utility function can be estimated. However, there are also some drawbacks. We have to deal with a trade-off between the number of levels and the complexity of the experiment as the combination of levels and attributes results in the total number of possible alternatives. This means that the number of alternatives increases exponentially with the number of attribute levels.

Some aspects regarding the choice of attribute levels are important to consider. First of all, they should be sufficiently wide to make respondents indeed "jump" between the proposed alternatives. Also, it should be possible to contrast an increase in the level of one attribute with a decline in the level of another attribute. That is, respondents should be forced to overcome trade-offs (Bateman et al. 2002; Telser 2002). To obtain meaningful results the attribute levels not only have to be plausible and realistic but also have to reflect the broad range of individuals' preferences. Besides these requirements, a starting point for defining adequate attribute levels is to "bracket" existing values i.e. in most cases the status quo.

2.2.3 Applying an Experimental Design

In the third step, the design of the DCE has to be considered. The *complete factorial design* (CFD) is the one containing all possible combinations of attributes and their levels. The great advantage of the CFD is its statistical property. It ensures that all main effects of attributes are independent from each other and attribute levels occur with the same frequency. These are the properties of an orthogonal and balanced design (Zwerina 1977). Per definition, these designs are 100 % efficient (Kuhfeld 2006). However, in reality the CFD results in a huge number of alternatives that cannot be realized in an experiment. Thus, the design, i.e. the number of alternatives, needs to be reduced. Although there again exists no strong advice how many decisions respondents can handle in an experiment, as a rule of thumb past DCEs have on average used 12 decisions per respondent (De Bekker-Grob et al. 2012).

The next question to answer is how the CFD should be reduced. The aim is to draw a sample out of the CFD which stands representatively for the CFD and limits the loss of information. The so called *fractional factorial designs* (FFD) (Street et al. 2008) can be obtained by using different mathematical algorithms. The most common form of a FFD is the *orthogonal main effects design* (Street et al. 2008). However, this design is very restrictive if some combinations of attribute levels are not realistic or mutually exclusive (Kuhfeld 2006; Kuhfeld et al. 1994). Given these shortcomings, *optimal designs* have grown in popularity

(Kuhfeld et al. 1994; Kanninen 2002). Optimal designs aim at combining attribute levels in a form that they maximize the information from the choice behavior of the respondents. The efficiency of optimal designs is mostly evaluated by the *D-efficiency* criterion.[5]

2.2.4 Generating and Visualizing Choice Sets

After reducing the number of alternatives using an experimental design strategy, one has to decide how to present these to the respondents. Usually, the attributes are visually presented in a table. The advantage is that respondents are able to directly compare the levels of the attributes *row by row*. Another possibility is to use diagrams for visualization and to make the substitutive character and the trade-offs explicit (for an example see Pfarr 2013). The aim of visualizing the choice sets should be to simplify the decision situation for the respondents and to avoid additional complexity.

Next, the number of alternatives each respondent has to compare in one situation has to be defined. In most cases, binary choice situations were used (Guttman et al. 2009). However, in 30 % of the DCEs respondents were faced with a multinomial choice situation. How many choice tasks individuals can handle depends on the complexity as well as on the number of attributes and levels. Thus,

> the fewer the number of attributes and levels, the more the number of choice tasks that can be allotted to each person (Bateman et al. 2002).

Further, for unbiased estimates it could be necessary that all individuals have similar knowledge about the current status quo or a clear understanding of the true state. Therefore, respondents should be provided with detailed instructions and a description of the choice process as well as the attributes and their possible realizations. Kjær (2005) suggests to incorporate examples of the choice task within the introduction to avoid a learning process during the first decisions.

2.2.5 Collecting Data

Finally, the last step in preparing a DCE is to administer the survey. In this context, the method of collecting the data must be chosen (Amaya-Amaya et al. 2008). This comprises the definition of (i) the population, which is dependent on your goals, (ii) the sample procedure, i.e. a simple random sample or a stratified quota sample, (iii) the sample size, which is contingent upon the heterogeneity of the population and the precision to which parameters should be estimated as well as (iv) the interview technique, i.e. for example face-to-face, telephone or online surveys

[5] It is beyond the scope of the contribution to discuss all existing experimental designs in detail. For more information regarding experimental design refer to Zwerina (1997) and Street et al. (2008).

(Louviere et al. 2000; Ben-Akiva and Lermann 1985). While telephone surveys are comparatively cheap, complex scenarios with a number of attributes are difficult to describe. Participants may have difficulties to grasp all aspects of the alternatives. Sending additional information via mail may help to mitigate the problem but it is difficult to control how the participants make use of this information. Online surveys allow for graphical support, but there are still parts of the population that have less access to the internet which might have a negative effect on the representativeness of the sample. The gold standard—but usually also the most expensive option—are still face-to-face interviews with experienced interviewers who can also ensure that all participants take their decisions by themselves and are not influenced by others, which is a potential downside of online and mail surveys.

Before the survey is conducted, the design and experimental setup should be evaluated in a pre-test to uncover possible problems and errors.

3 Eliciting Preferences in Health Care

Assessing patients' preferences to design optimal health care systems is a challenging task. Understanding their needs and tailoring supply of care to these particular requirements may help increasing the public satisfaction with health care delivery and health insurance. Patient mobility is a crucial parameter in this respect. Are patients willing to travel for certain services? If so—how far? To which extent do subgroups of patients differ regarding their capability to travel? This is not only important with regard to the choice of a health care provider within the patient's country. In many cases—and not only in boarder regions—cross-border care is gaining relevance. Any changes in the benefits catalog or in supply structures may induce people to vote over these modifications with their feet, resulting in changing demand for certain providers or even health care systems. Thus, knowledge about preferences may save regulators and providers from decisions with unintended side effects.

We do not aim at providing a systematic review of the literature but rather to give the reader an impression of the broad range of opportunities provided by a DCE. For a comprehensive account of the health care related DCE literature we recommend the systematic review provided by De Bekker-Grob et al. (2012). They identify 114 studies published in English between 2001 and 2008 which use DCEs related to health care. Seven categories, i.e.

- patient or consumer experience factors;
- valuing health outcomes;
- investigating trade-offs between health outcomes and patient or consumer experience factors;
- estimating utility weights within the QALY framework;
- job-choices;
- developing priority setting frameworks;

- health professional's preferences for treatment or screening options for patients;
- and other studies.

are used to systematize the literature, thereby underlining the great flexibility and wide applicability of this method.

To illustrate some of the capabilities, but also some of the limitations of DCEs in health economics we pick three more narrowly defined topics and present examples of DCEs for health insurance design, for the demand for pharmaceuticals and for provider choice.

3.1 Preferences for Health Insurance

In health economics, the demand for health insurance and the optimal design of health insurance contracts have been discussed for a long time (Short and Taylor 1989). While theoretical approaches often analyze incentive compatibility, most empirical work focuses on causal relationships e.g. between socio-economic characteristics and the choice of health plans using revealed preference data. As explained above, DCEs expand the scope and give insights into customers' preferences of existing and hypothetical health insurance setups. Both aspects, the financing component, i.e. the design of the insurance tariff, was well as the benefits component, i.e. the coverage of the insurance plan or social health insurance system can be jointly analyzed. These characteristics of DCEs become especially important when health reform options for countries with social health insurance systems are discussed. Most of these systems struggle with financing their comprehensive benefits. So both sides may need to be revised: Opening up new sources of financing, e.g. by co-payments, and restricting the coverage. To get an idea of how citizens (and thus potential voters) may react to such changes, DCEs are a valuable source of information. For example, the willingness to pay (WTP) for certain components of the benefits catalog could be estimated and the sensitivity to the framing of price signals could be tested.

The study by Pfarr and Schmid (2013) aims at the general health insurance setup that needs to be decided about in a political process—do citizens prefer social health insurance over private health insurance? And if so, what are the determinants? The authors link their DCE to a basic public choice model and find a strong preference for social health insurance even under the well-off parts of the population. Prior studies focusing on similar topics such as public redistribution also try to capture preferences. However, these studies rely on survey questions asking participants for their opinion on questions like "Should the government spend much more, more, a little bit more, just the same, a little bit less or much less for the poor/sick/unemployed?" (Alesina and Giuliano 2011; Corneo and Gruner 2000). Such questions are prone to bias (e.g. social desirability or excessive generosity). The DCE has the virtue to confront participants with the costs of their decisions—such reducing potentially excessive generosity—and forces participants to overcome trade-offs between alternative uses of a limited budget. The study by Pfarr and Schmid is the first to apply a DCE on this particular question.

A more typical study is the one conducted by Gryd-Hansen and Slothouus (2002). The authors investigate preferences for enhancing the benefits catalog in the tax financed Danish health care system using a representative sample of the Danish population. They make three pairwise comparisons with the status quo as reference. They find that including medical innovations as well as screening, preventive services and free choice of hospital seems to enhance the utility for the respondents. With a look at the financing aspect, individuals prefer co-payments over higher taxes or contributions. Telser et al. (2004) look at cost containment in the Swiss health care system and whether this leads to losses in utility. Also building on a representative sample of the population, they find that constraining the free choice of physician goes along with high utility losses and that the perceived restrictions increase with age and income. However, enhancing the benefits catalog by including aspects of long-term care financed by higher co-payments for those aged 50 and above is refused. The study by Becker and Zweifel (2008) analyzes the relationship of age and willingness to pay for additional services. They hypothesize that, if preferences of patients are heterogeneous and thus differ from a one size fits all uniform benefits package, introducing choice among different health insurance contracts would be welfare enhancing. Attributes used in the experiments are related to cost sharing and premium height, alternative medicine and medication by generics and the access to innovation. They use a survey of 1,000 people above the age of 24 in the German and French-speaking parts of Switzerland and find a large diversity of preferences. Older patients prefer the status quo over alternatives. Introducing high deductibles or cost-sharing would require premium reductions to compensate for utility losses.

Similar studies investigate cross country differences regarding preferences for components of the respective benefits catalog. All of the following studies are based on two experiments in Germany and the Netherlands. Böcken et al. (2007) focus on acceptance and utilization of hypothetical social health insurance systems and find similar attitudes in both countries. Nevertheless, there exist differences in the demand for health care due to existing characteristics of the national health care systems. For example, in the Netherlands, specialists are located at hospitals, while in Germany, specialist services are provided in the ambulatory care sector. This effects health care utilization and may confound the results. People in Germany object increases in contributions and have a strong preference for free choice of physician. In the Netherlands, respondents consider free choice of doctors and a guaranteed maximum waiting time for hospital care in selected hospitals as utility improving. Furthermore, the Dutch population shows higher price sensitivity and a higher willingness to switch their health insurance company compared to Germans. Zweifel et al. (2010) focus on differences in preferences for reforms of the health care systems and investigate the impact of additional information to reduce the status quo bias. The authors find the same attributes to have a positive or negative willingness to pay for both countries but differences in their magnitude.

According to the authors this could be interpreted as difference in the preference structure. In addition, consumer information in the Netherlands seems to lead to more conscious decisions about different characteristics of attributes and their consequences. The last study by MacNeil-Vroomen and Zweifel (2011) aims at measuring differences in preferences for health insurance of individuals with and without chronic conditions in Germany and the Netherlands. They hypothesize that in both countries differences between the subgroups exist and test for the impact of chronic conditions. They find that patients with a chronic illness tend to object restrictions of the access to physicians more. The authors conclude that individuals need to be highly compensated e.g. to accept the participation in a disease management program which implies restrictions on physician choice.

Another national study which focuses on Germany is offered by Streibelt and Bethge (2013). They investigate patients' preferences for integrated care programs and conduct a DCE with patients in need for a total knee or hip replacement surgery. In detail, they look at preferences for the following four attributes: treatment consultation, type of post-operative treatment, cooperation of hospital and rehabilitation clinic and the amount of additional payment. With 340 patients participating in this study, they found all attributes showing a significant impact on the decision to treat. Patients show a higher benefit if an integrated care program includes cooperation of hospitals and rehabilitation clinic, face-to-face consultation by a medical doctor and structures post-operative care.

To conclude, some findings on preferences for health insurance seem to be confirmed by various DCEs. Consumers are price sensitive and are very reluctant to accept restrictions to their benefits packages. However, especially the cross country comparisons illustrate some of the pitfalls every researcher should be aware of. Most importantly: No matter how good the introduction to the DCE is, participants are always influenced by the experiences they have made in the past and thus by the health care system that they experience day to day. A very careful wording of the scenarios and an adequate econometric specification are paramount to avoid severe bias and underline the necessity of comprehensive pre-tests.

3.2 Preferences for Pharmaceuticals

While the first part dealt with choice of health insurance plans and the design of social health insurance systems, studies on preferences for pharmaceuticals have great importance when investigating new pharmaceutical treatment methods and conducing cost-utility evaluation. Hence, many experiments concentrate on specific disease patterns. Among the few more general analyses with no focus on a special treatment form or disease pattern, Merino-Castelló (2003) studies the consequences of supplier inducement and brand loyalty on the decision to purchase drugs. In a world with better access to information, patients have the opportunity to question the physician's advice. The experiment focuses on two groups of drugs, Amoxiciline for common infections (throat infection) and Statins for chronic diseases (high blood cholesterol). Attributes comprise three groups: brand loyalty represented by commercial name of the drug and reputation of the laboratory (knowledge about the producer), price elasticity and whether the drug was purchased after physician's or pharmacist's recommendation as attributes of supplier inducement. Using a stated preference approach the study shows a significant importance of supplier inducement. For individuals with chronic diseases experts' inducement plays a more important role than brand loyalty. Moreover, patients tend to trust more in a physician's prescription and a brand name with rising age.

Most studies analyze patients' preferences for a medication or treatment of a specific illness. Ratcliffe et al. (2002) explore patients' preferences for homeopathic and conventional asthma treatment using a conjoint analysis with 300 randomly assigned patients of two large London hospitals. They find distinct differences in the structure of patients regarding age, co-determination and additional treatments. The attribute "extent to which the doctor treated the patient as a whole person" was statistically significant only for those treated homeopathically. The preference for this attribute further increased with age. Preferences for managing asthma are analyzed by King et al. (2007) using a DCE embedded in a randomized clinical trial with patients suffering from mild-moderate persistent asthma. The respondents were asked after 6 weeks' treatment with one of two pharmaceuticals whether first to continue with the medication, second to change to a hypothetical treatment alternative and third to take no preventive medicine. Attributes encompassed costs of medication, frequency to take the drug, daily activities, doctor recommendation and side effects among others. The results indicate a preference for the status quo depending on the kind of medication. Respondents show clear preferences for the ability to participate in usual daily activities and sport, for minimal symptom and for medications without side-effects. De Bekker-Grob et al. (2008, 2009) analyze preferences for osteoporosis treatment in two studies. In the first one, they explore women's preferences with high fracture risk who may benefit from a preventive drug treatment on the attributes effectiveness, side effects, total treatment duration, out of pocket costs and route of drug administration (De Bekker-Grob et al. 2009). All of these showed a significant impact for the women's choices. In addition, respondents preferred taking preventive drugs disregarded of possible side effects and out-of pocket payments. In the second analysis, De Bekker-Grob et al. (2008) compare preferences of general practitioners and patients for a preventive osteoporosis drug treatment. Fourty GPs and 120 women aged above 60 from the Rotterdam area participated in the DCE. Attributes cover the route of drug administration (tablet or injection), effectiveness of treatment (10-year risk reduction of hip fracture), adverse effects (nausea), treatment duration and total cost for patients. While all attributes are important for both groups, GPs prefer a preventive drug treatment of osteoporosis less than patients do. Instead, they place a higher weight on effectiveness and short total preventive treatment duration. The question of funding orphan drugs is at the core of the study by Mentzakis et al. (2011). University students are asked to state their preferences regarding public funding of orphan drugs versus drugs for common diseases on the following attributes: frequency of disease, cost of single patient treatment, budget impact (total cost of funding), severity of disease without treatment and impact of treatment on patient's health. The main result is that respondents have no preferences to spend money from the government for drugs on rare diseases. There seems to be no difference in the WTP per life year between rare and common diseases. The same applies to attributes of the coverage decision like costs, disease severity and treatment effectiveness. A last study by Sennhauser and Zweifel (2010) assesses preferences for new therapies and deals with the question of whether to include new pharmaceuticals in the benefits catalog. The DCE was conducted in Germany on preferences for modern insulin treatment (analog insulin vs. NPH insulin). Two attributes are focused on: co-payments and increased SHI contributions. Results show differing preferences between the subgroups. Non-diabetics and insulin naïve type 2 diabetics prefer to include the new medication through co-payments whereas those affected type 1 and insulin-treated type 2 diabetics prefer increased contributions. The authors argue that the inclusion of the new insulin drugs is justified as the WTP exceeds the extra treatment costs.

A different field of pharmaceuticals is analyzed in a systematic review by Naik-Panvelkar et al. (2013). They focus on DCEs with pharmacy background for the years 2005–2010, all published in English. They identified 12 studies from which only six deal with patients' preferences and two more analyze both patients' and pharmacists' preferences. Most of the literature focused on process and provider attributes and only few on health-outcome attributes. In general, they find that monetary attributes are considered to be very important for both patients and pharmacists and tendency to favor current pharmacy or pharmacy services (status quo bias).

To sum up, studies on pharmaceuticals give insight into patients' preferences for treatment alternatives, their willingness to pay or the financial burdens they would take to receive such a treatment. Moreover, possible side-effects of treatment play an important role for the drug treatment decision. The broadband of the studies presented here shows also some shortcomings. There exists no universal design of DCEs in the pharmaceutical context. Even in the same health care system, experiments may differ between patient groups and drugs. Nevertheless, they give some valuable insights into attributes important for patients' and physicians' preferences on drug treatments.

3.3 Preferences for Providers

While the studies regarding preferences for health insurance and preferences for pharmaceuticals give interesting insights into the capability of DCEs, the aspect of patient mobility is of no particular importance. This changes considerably when we turn towards patients' preferences regarding their choice of provider. It is intuitive that the distances between a patient and different alternative providers is a critical component in the decision making process. The patient's valuation of the attribute "distance" reflects to a large extent his mobility.

One of the very first pilot studies following an approach similar to a DCE[6] was published in 1976 by Wind and Spitz (1976). The authors use a small sample from the Philadelphia suburban area and differentiate the proximity of a hospital in the three categories "Downtown area with easy parking and access", "Downtown are with difficult access and parking" and "in your neighborhood". Among several significant characteristics they find that proximity is the most important one.

Later studies have a broader regional or national scope aiming for representative results and implemented proximity by distance, travel time and/or travel costs. Two studies from the UK provide extensive documentation of such DCEs. Both are also good examples for the potential of DCEs to predict patient behavior when health reform is about to open new choice for patients, e.g. by changing rules for referrals. The first by Burge et al. (2005) focus on patients on waiting lists who are offered alternatives for quicker treatment. The main interest of the study was in the decisive factors that would drive the patients' decisions and in the relative weight of these factors. One of their findings is that for "every additional hour of travel, on average, patients would require a reduction in waiting time of 2.1 months to take up the offer of an alternative hospital" (p. 58). The negative appreciation of distance was even stronger when traveling abroad was offered as an alternative. Even when being compensated for costs of traveling the waiting time needed to be reduced by "around 5.4 months at the alternative hospital before accepting the offer to travel abroad for treatment" (p. 57). The results also indicate that depending on socio-demographic characteristics such as income and age the willingness to travel varies considerably. Another interesting aspect of this study is that the authors used both stated and revealed preferences data. The second study by Burge et al. (2006) looks at patients at the point of referral, i.e. before they eventually enter a waiting list. When introducing new choice options for patients at the point of referral policy makers were interested in two questions: Which pieces of information would be taken into consideration by the patients and to which extent would each piece of information influence the decision making process also in relation to other factors like travel time? Focusing on their results regarding the latter aspect it is interesting to note that across all modes of transportation the disutility of traveling is strongest for the first hour of traveling. Furthermore the authors find that patients with income of more than £30,000 have a lower disutility per additional unit of travel time than those with lower incomes. Other studies on preferences for providers omit attributes that capture distance. This may well be reasonable when this aspect is not of interest for the analysis. For example, Hjelmgren and Anell (2007) analyze the population preference for the choice among two different types of primary care models for Sweden. The key question is about attributes that influence individual choice between family physicians

[6] The authors do not apply a DCE, but use conjoint measurement. However, the general notion is related to DCEs.

(general practitioners) and primary care teams, consisting of physicians and nurses. The DCE was conducted among 1,600 individuals with a response rate of 58 %. Older individuals and individuals in poor health preferred registering with a GP whereas working individuals and those living in regions distant from hospitals chose to register with a primary care team. Similarly, Hanson et al. (2005) analyze factors influencing the demand for hospital care in Zambia and the perceived quality. As no revealed preference data of sufficient quality was available, collecting stated preferences data by conducting a DCE was the only viable option. They focus on two acute medical conditions: cerebral malaria for adults and acute pneumonia for children. They find that technical quality, i.e. the thoroughness of the physical exam, is the most important quality attribute. Others are staff attitudes and drug availability. As the aim of the study was to gain a better understanding of quality perceptions in a developing country and not to predict hospital utilization it seems reasonable that distance was not accounted for.

Using DCEs to capture patients' preferences for providers remains a highly appreciated method, also to inform providers. Just looking at ongoing projects in Germany, Koppe et al. (2012) published evidence from a pilot study that focuses on preferences of patients regarding hospital treatment. Such information may be useful in a hospital market with a higher degree of competition to enhance a hospital's customer orientation. With 56 individuals in this pilot, the authors conclude that medical quality, coordination and integration of treatment, cleanliness, time for care and amiability of staff are of major importance. Unexpectedly, the study suggests that the distance between residence and hospital is not an important attribute. However, results based on such pilot studies have to be treated with caution as sample selection and the number of participants do not allow for robust estimates.

As a last aspect it should be noted that discrete choice experiments also allow to derive welfare estimates. So far, most examples refer to analyses with revealed preference data. For example, Tay (2003) predicts hospital choice on the basis of revealed preference data and estimates welfare losses due to potential hospital closures. Intuitively the local community is affected most and has to suffer disutility from increased travel times. Discrete choice experiments are feasible options in cases in which such revealed preference data are not available. This can be due to a general lack of data or due to characteristics (i.e. attributes) of interest that are likely to influence future patient behavior but are not yet reflected in current data as these characteristics are yet to be implemented (i.e. for example by establishing an innovative form of health-IT driven portal clinic in a rural area).

As the presented studies show, DCEs provide a wide range of options for patients' preferences regarding providers. Patient mobility can be assessed and the effects of different socio-demographic characteristics can be accounted for. DCEs can be a valuable tool for making better informed decisions when making decisions no matter if on provider or government level.

4 Concluding Remarks

Eliciting patients' preferences is central for complex decisions in health economics. Will the patients be willing to travel to an alternative location of a newly built hospital? Will they appreciate new options of cross border care? However, such preferences are usually latent constructs. Especially when new products, services or processes are to be designed, no data on past decisions are available. In such a context discrete choice experiments provide a valuable tool to collect stated preferences data.

As outlined in the theory part of this article, DCEs have certain properties that make them especially helpful to researchers. A well-constructed and implemented DCE will break down a complex decision situation in a way that makes it easy to understand for participants. By using qualitative and quantitative approaches relevant characteristics, i.e. attributes, can be identified and the respective levels can be assigned. At the same time, trade-offs as well as budget constraints can be implemented. This helps to avoid that participants exaggerate their true appreciation of certain aspects or understate their true willingness to pay for strategic reasons.

The literature cited in the sections on applied DCEs in the area of health insurance or pharmaceuticals provide evidence for the practical application of this approach. Numerous studies illustrate the range of questions that can be investigated by DCEs and the various methodological options that have been implemented in the past. The applicability of this method to the central theme of this volume, i.e. mobility, is underlined in the section on preferences for providers. Various aspects of mobility such as distance, travel time, comfort and costs can be implemented within one decision scenario. Depending on the research question a varying number of attributes can be used to adjust the precision with which this aspect is captured in a DCE.

However, every researcher should be aware that well-designed DCEs can provide substantial and detailed evidence, while poorly-designed DCEs are worthless. The necessary steps to plan and implement a DCE were only briefly outlined and can only provide a starting point for further information on this topic. The literature cited in this article should serve as a solid basis.

References

Alesina, A., & Giuliano, P. (2011). Preferences for redistribution. In J. Benhabib, M. O. Jackson, & A. Bisin (Eds.), *Handbook of social economics* (pp. 93–132). Amsterdam: North Holland.

Amaya-Amaya, M., Gerard, K., & Ryan, M. (2008). Discrete-choice experiments in a nutshell. In M. Ryan, K. Gerard, & M. Amaya-Amaya (Eds.), *Using discrete choice experiments to value health and health care* (pp. 13–46). Dordrecht: Springer.

Anderson, S. P., Palma, A., De, & Thisse, J. F. (1992). *Discrete choice theory of product differentiation*. Cambridge: MIT Press.

Bateman, I. J., Carson, R. T., Day, B., Hanemann, M., & Hanley, N. (2002). *Economic valuation with stated preference techniques: A manual*. Cheltenham: Elgar.

Becker, K., Zweifel, P., & Becker, K. (2008). Age and choice in health insurance: Evidence from Switzerland. *The Patient: Patient-Centered Outcomes Research, 1*(1), 27–40.

Ben-Akiva, M., & Lermann, S. R. (1985). Discrete choice analysis. *Theory and applications to travel demand*. MIT Press: Cambridge.

Böcken, J., Janus, K., & Schwenk, U. (2007). *Neue Versorgungsmodelle im Gesundheitswesen [New treatment models in health care]*. Gütersloh: Bertelsmann Stiftung.

Burge, P., Devlin, N., Appleby, J., Gallo, F., Nason, E., & Ling, T. (2006). Understanding patients' choices at the point of referral. *RAND Europe technical report*, TR-359-DOH. RAND Cooperation: Santa Monica.

Burge, P., Devlin, N., Appleby, J., Rohr, C., & Grant J. (2005). London patient choice project evaluation. A model of patients' choices of hospital from stated and revealed preference choice data. *RAND Europe technical report*, TR-230. RAND Cooperation: Santa Monica.

Cameron, A. C., & Trivedi, P. K. (2009). *Microeconometrics using Stata*. College Station: Stata Press.

Corneo, G., & Grüner, H. P. (2000). Social limits to redistribution. *American Economic Review, 90*(5), 1491–1507.

De Bekker-Grob, E. W., De Ryan, M., & Gerard, K. (2012). Discrete choice experiments in health economics: A review of the literature. *Health Economics, 21*(2), 145–172.

De Bekker-Grob, E. W., De Essink-Bot, M., Meerding, W. J., Koes, B. W., & Steyerberg, E. W. (2009). Preferences of GPs and patients for preventive osteoporosis drug treatment: A discrete-choice experiment. *Pharmacoeconomics, 27*(3), 211–219.

De Bekker-Grob, E. W., De Essink-Bot, M., Meerding, W. J., Pols, H. A. P., Koes, B. W., & Steyerberg, E. W. (2008). Patients' preferences for osteoporosis drug treatment: A discrete choice experiment. *Osteoporosis International, 19*(7), 1029–1037.

Guttmann, R., Castle, R., & Fiebig, D. G. (2009). Use of discrete choice experiments in health economics. An update of the literature book. *CHERE Working Paper*. Sydney: University of Technology.

Gyrd-Hansen, D., & Skjoldborg, U. S. (2008). The price proxy in discrete choice experiments: issues of relevance for future research. In M. Ryan, K. Gerard, & M. Amaya-Amaya (Eds.), *Using discrete choice experiments to value health and health care* (pp. 175–193). Dordrecht: Springer.

Gyrd-Hansen, D., & Slothuus, U. (2002). The citizen's preferences for financing public health care: A Danish survey. *International Journal of Health Care Finance and Economics, 2*(1), 25–36.

Hanemann, M. (1983). Marginal welfare measures for discrete choice models. *Economics Letters, 13*(1), 129–136.

Hanson, K., McPake, B., Nakamba, P., & Archard, L. (2005). Preferences for hospital quality in Zambia: Results from a discrete choice experiment. *Health Economics, 14*(7), 687–701.

Hjelmgren, J., & Anell, A. (2007). Population preferences and choice of primary care models: A discrete choice experiment in Sweden. *Health Policy, 83*(2–3), 314–322.

Johannesson, M. (1996). *Theory and methods of economic evaluation of health care*. Dordrecht: Kluwer Academic Publishers.

Johnson, R. F., Mohamed, A. F., Özdemir, S., Marshall, D. A., & Phillips, K. A. (2011). How does cost matter in health-care discrete-choice experiments? *Health Economics, 20*(3), 323–330.

Kanninen, B. J. (2002). Optimal design for multinomial choice experiments. *Journal of Marketing Research, 39*(2), 214–227.

King, M. T., Hall, J., Lancsar, E., Fiebig, D., Hossain, I., Louviere, J., et al. (2007). Patient preferences for managing asthma: Results from a discrete choice experiment. *Health Economics, 16*(7), 703–717.

Kjær, T. (2005). A review of the discrete choice experiment—with emphasis on its application in health care. *Health economics paper*. Denmark: University of Southern.

Kløjgaard, M. E., Bech, M., & Søgaard, R. (2012). Designing a stated choice experiment: the value of a qualitative process. *Journal of Choice Modelling, 5*(2), 1–18.

Koppe, K., Bethge, S., & Mühlbacher, A. C. (2012). Präferenzmessung im Krankenhaus - wissen, was den Patienten wirklich wichtig ist [Measuring preferences for hospitals—what is really important for patients?]. *Gesundheitsökonomie und Qualitätsmanagement, 17*(3), 138–143.

Kuhfeld, W. F. (2006). Construction of efficient designs for discrete-choice-experiments. In: R. Grover & M. Vriens (Eds.), *The handbook of marketing research. Uses, misuses, and future advances* (pp. 312–363). Thousand Oaks: Sage Publications.

Kuhfeld, W. F., Tobias, R. D., & Garratt, M. (1994). Efficient experimental design with marketing research applications. *Journal of Marketing Research, 31*, 545–557.

Lancaster, K. J. (1966). A new approach to consumer theory. *Journal of Political Economy, 74*(2), 132–157.

Lancaster, K. J. (1971). *Consumer demand. A new approach*. New York: Columbia University Press.

Lancsar, E., Louviere, J. J., & Flynn, T. N. (2007). Several methods to investigate relative attribute impact in stated preference experiments. *Social Science and Medicine, 64*(8), 1738–1753.

Louviere, J. J., & Hensher, D. A. (1982). Design and analysis of simulated choice or allocation experiments in travel choice modelling. *Transportation Research Record, 890*, 11–17.

Louviere, J. J., & Street, D. J. (2000). Stated-preference methods. In D. A. Hensher & K. Button (Eds.), *Handbook of transport modelling* (1st ed., pp. 131–143). New York: Pergamon.

Louviere, J. J., & Woodworth, G. (1983). Design and analysis of simulated consumer choice or allocation experiments: an approach based on aggregated data. *Journal of Marketing Research, 20*(4), 350–367.

Louviere, J. J., Flynn, T. N., & Carson, R. T. (2010). Discrete choice experiments are not conjoint analysis. *Journal of Choice Modelling, 3*(3), 57–72.

Louviere, J. J., Hensher, D. A., & Swait, J. D. (2000). *Stated choice methods. Analysis and application*. Cambridge: Cambridge University Press.

MacNeil Vroomen, J., & Zweifel, P. (2011). Preferences for health insurance and health status. Does it matter whether you are Dutch or German? *The European Journal of Health Economics*, 12(1), 87–95.

Manski, C. F. (1977). The structure of random utility models. *Theory and Decision, 8*, 229–254.

Manski, C. F., & Lerman, S. R. (1977). The estimation of choice probabilities from choice based samples. *Econometrica, 45*(8), 1977–1988.

McFadden, D. (1974). Conditional logit analysis of quantitative choice behavior. In P. Zarembka (Ed.), *Frontiers in econometrics* (pp. 105–142). New York: Academic Press.

McFadden, D. (1981). Econometric models of probabilistic choice. In C. F. Manski & D. McFadden (Eds.), *Structural analysis of discrete data with econometric applications* (pp. 198–272). Cambridge: MIT Press.

Mentzakis, E., Stefanowska, P., & Hurley, J. (2011). A discrete choice experiment investigating preferences for funding drugs used to treat orphan diseases. An exploratory study. *Health Economics, Policy and Law, 6*(3), 405–433.

Merino-Castelló, A. (2003). Demand for pharmaceutical drugs: A choice modelling experiment. *UPF Economics and Business Working Paper* No. 704, Departament d'Economics i Empresa, Universitat Pompeu Fabra.

Mitchell, R. C., & Carson, R. T. (1989). *Using surveys to value public goods. The contingent valuation method* (4th ed.). Washington: Johns Hopkins University Press.

Naik-Panvelkar, P., Armour, C., & Saini, B. (2013). Discrete choice experiments in pharmacy: A review of the literature. *International Journal of Pharmacy Practice, 21*(1), 3–19.

Pfarr C., & Schmid, A. (2013). The political economics of social health insurance: The tricky case of individuals' preferences. *Discussion Paper 01–13*, University of Bayreuth.

Pfarr, C. (2012). Meltzer-Richard and social mobility hypothesis: Revisiting the income-redistribution nexus using German choice data. *Discussion Paper 10–12*, University of Bayreuth.

Pfarr, C. (2013). Einkommen, Mobilität und individuelle Präferenzen für Umverteilung. Ein Discrete-Choice-Experiment. Dissertation. Beiträge zur Finanzwissenschaft, vol. 30. Mohr Siebeck, Tübingen.

Ratcliffe, J., van Haselen, R., Buxton, M., Hardy, K., Colehan, J., & Partridge, M. (2002). Assessing patients' preferences for characteristics associated with homeopathic and conventional treatment of asthma: A conjoint analysis study. *Thorax, 57*(6), 503–508.

Samuelson, P. A. (1938). A note on the pure theory of consumer's behaviour. *Economica, 5*(17), 61–71.

Sennhauser, M., & Zweifel, P. (2010). A pharmaceutical innovation. Is it worth the money? Whose money? *Working Paper No. 0914*, Socioeconomic Institute, University of Zurich.

Short, P. F., & Taylor, A. K. (1989). Premiums, benefits and employee choice of health insurance options. *Journal of Health Economics, 8*(3), 293–311.

Street, D. J., Burgess, L., Viney, R., & Louviere, J. J. (2008). Designing discrete choice experiments for health care. In M. Ryan, K. Gerard, & M. Amaya-Amaya (Eds.), *Using discrete choice experiments to value health and health care* (pp. 47–72). Dordrecht: Springer.

Streibelt, M., & Bethge, M. (2013). Patientenpräferenzen bei der Ausgestaltung integrierter Vorsorgungsprogramme: ein Discrete Choice Experiment [Patient preferences for integrated care programmes: a discrete choice experiment]. *Gesundheitsökonomie und Qualitätsmanagement, 15*(6), 288–295.

Tay, A. (2003). Assessing competition in hospital care markets: The importance of accounting for quality differentiation. *RAND Journal of Economics, 34*, 786–814.

Telser, H. (2002). *Nutzenmessung im Gesundheitswesen*. Kovac, Hamburg: Die Methode der Discrete-Choice-Experimente.

Telser, H., Vaterlaus, S., Zweifel, P., & Eugster, P. (2004). *Was leistet unser Gesundheitswesen? [What our health care system can afford]*. Zurich: Ruegger.

Train, K. E. (2009). *Discrete choice methods with simulation* (2nd ed.). New York: Cambridge University Press.

Wind, Y., & Spitz, L. K. (1976). Analytical approach to marketing decisions in health-care organizations. *Operations Research, 24*(5), 973–990.

Zweifel, P., Leukert, K., & Berner, S. (2010). Preferences for health insurance in Germany and the Netherlands: A tale of two countries. *Working Paper No. 1002*, Socioeconomic Institute, University of Zurich.

Zwerina, K. (1997). *Discrete choice experiments in marketing. Use of priors in efficient choice designs and their application to individual preference measurement*. Heidelberg: Physica-Verlag.

Implications of the EU Patients' Rights Directive in Cross-Border Healthcare on the German Sickness Fund System

Mathias Kifmann and Caroline Wagner

Abstract We examine the implications of the EU directive on the application of patients' rights in cross-border healthcare on the German sickness fund system. Since Germany implemented most requirements of the directive already in 2004, we first review Germany's experience with EU cross-border healthcare. We then focus on the possible effects of increased EU cross-border healthcare. While this gives patients more choice, the German sickness fund system faces a number of challenges. EU cross-border care may undermine efforts to keep healthcare expenditure under control. Cross-border care can also increase inequality of access. Furthermore, promoting cross-border care can be a means for sickness funds to attract good risks. We discuss these challenges and point out possible policy responses.

Keywords EU patients' right directive · EU cross-border healthcare · Physician payment · Expenditure control

JEL Classification F15 · H51 · I13 · I18

M. Kifmann (✉)
Universität Hamburg, Hamburg Center for Health Economics, Esplanade 36 20354 Hamburg, Germany
e-mail: mathias.kifmann@wiso.uni-hamburg.de

C. Wagner
Wissenschaftliches Institut der Techniker Krankenkasse für Nutzen und Effizienz im Gesundheitswesen (WINEG), (Scientific Institute of Techniker Krankenkasse for Benefit and Efficiency in Health Care), Bramfelder Straße 140 22305 Hamburg, Germany
e-mail: dr.caroline.wagner@wineg.de

R. Levaggi and M. Montefiori (eds.), *Health Care Provision and Patient Mobility*, Developments in Health Economics and Public Policy 12,
DOI: 10.1007/978-88-470-5480-6_3,

1 Introduction

By October 2013, all EU member states are required to implement the EU patients' rights directive in cross-border healthcare. After the European Court of Justice Kohll and Decker decisions in 1998, it is the second milestone in facilitating medical treatment in a member state different from the state where patients are insured. With few exceptions, individuals now have the right to obtain treatment in other member states if they have the right to this treatment in their home country. The healthcare system which covers the individual must reimburse up to the amount which it would have paid for the same treatment in the home country.

From the point of view of patients, the EU patients' rights directive offers new opportunities. Quality of treatment may be perceived as higher in other member states. Also waiting times may be shorter. Furthermore, if treatment in the home country involves considerable copayments, traveling to other member states can be financially attractive. Increased patients' mobility can also encourage competition between healthcare providers leading to higher quality and lower costs in the home country.

However, the EU patients' rights directive also poses a challenge to national healthcare systems. National systems of treatment allocation may be undermined as in the case of waiting lists. In particular, it may be regarded as unjust when those who can afford to travel to other member states get better access to healthcare. This also holds for national systems of planning and financing which need to be revised when a considerable number of patients obtain their treatment in other member states. The directive here allows member states to ask for prior authorization which can be refused if treatment can be provided in the home country within a time limit which can be medically justified.

Furthermore, national systems of expenditure control are challenged by the EU directive. At first sight, the procedure for reimbursing cross-border care appears to have no impact on expenditure or even help to reduce expenditure because the amount of the reimbursement should not be larger than in the home country and must not exceed actual costs of care. Yet, it can be costly to determine this amount since the corresponding treatment and its reimbursement in the home country must be clarified. In particular, the latter may prove difficult unless the country uses a fee-for-service system. Many states, however, use budgets or put caps on the amount of reimbursement which causes nominal fees to be higher than what is effectively reimbursed.

In countries with competing health insurers like Germany, cross-border care also raises the issue of risk selection. Specifically, users of planned cross-border care may on average be good risks since they are still able to travel. Health insurers may therefore excessively promote cross-border care. This can be at the expense of those who are dependent on treatment in the home country.

For Germany, the implementation of the EU patients' rights directive required only a minor change in law—the formation of a national contact point for cross-border healthcare. All other requirements were already met by the Health

Modernization Act of 2004 ("Gesetz zur Modernisierung der gesetzlichen Krankenversicherung") which allowed members of sickness funds to obtain treatment in all EU member states and the four members of the European Free Trade Association. Treatment costs are reimbursed up to the amount paid for the same treatment in Germany.

For outward patient mobility, Germany therefore provides an interesting case study for possible impacts of the directive. In our contribution, we discuss the German experience with patient mobility. We present the results of a survey commissioned by Techniker Krankenkasse, one of Germany's largest sickness funds. Furthermore, we point out the challenges posed to the German sickness fund system by cross-border healthcare and discuss policy implications.

We proceed as follows. In the following Sect. 2, we describe the German sickness fund system and the legal framework for cross-border healthcare. Section 3 presents data on Germany's experience with EU cross-border healthcare. Possible effects of increased EU cross-border healthcare for the German sickness fund system are analyzed in Sect. 4. Building on our results, we discuss possible policy implications in Sect. 5. A final section concludes.

2 The German Sickness Fund System and Cross-Border Healthcare

2.1 The German Sickness Fund System

In Germany, about 88 % of the population is insured by the social health insurance (SHI) system ("Gesetzliche Krankenversicherung"). About 12 % of the population obtains coverage through the private health insurance system (PHI, "Private Krankenversicherung"). These are mainly individuals who opted out of the SHI system and civil servants. The SHI system is largely financed by the contributions which are shared by employees and their employers unless members are self-employed. The contribution rate is automatically deducted from monthly salaries. Currently, the universal contribution rate amounts to 15.5 %, with the members sharing 8.2 % and employers 7.3 % of the contribution rate. Total contributions of all members are collected in a central fund ("Gesundheitsfonds") and distributed to sickness funds in form of risk-adjusted capitations.[1]

The remuneration of the health service providers is negotiated in complex corporatist social bargaining processes between SHI funds and provider organizations mainly on the state level ("Bundesländer"). For outpatient care, sickness funds pay a capitation to physicians' associations to cover most expenses. These

[1] The German risk-adjustment scheme differentiates capitations according to 40 age and gender groups, 6 reduced earning capacity groups and 155 hierarchic morbidity groups which are based on 80 diseases.

associations distribute their total revenue to physicians according to a complex catalog which specifies fees for cases as well as for specific services. Physicians face a reduction in their fees once they have treated more than a predefined number of cases in a quarter. Hospital reimbursement is mainly based on Diagnostic Related Groups (DRGs). Similar to physicians' reimbursement, target budgets limit total expenditure. If a hospital treats more cases than agreed upon in negotiations, it faces reductions in its DRG payments for additional patients.

Presently, the SHI system consists of around 130 sickness funds. These are corporations under public law and have historically the right to self-government. The key principles are solidarity (everyone is covered and pays a contribution unrelated to the health status) and subsidiarity (local decision-making and personal responsibility). Sickness funds compete for SHI members with their services and with their cost management. In particular, they try to avoid that they have to charge additional premiums to cover their expenses.

SHI members can obtain treatment from physicians which belong to a physicians' association. Access to medical care is mainly free. Copayments exist for prescription drugs and for inpatient care. They cannot exceed 2 % of household income. Traditionally, general practitioners have no gatekeeping role[2] and patients are free to go directly to an SHI-affiliated specialist doctor of their choice. Importantly, members generally receive benefits in kind ("Sachleistungsprinzip"), implying that they are not billed for services.

2.2 *Regulation of Cross-Border Healthcare*

Until 1998, the Council Regulation (EEC) No. 1408/71[3] provided the sole basis for EU cross-border care. This regulation gave EU citizens the right to the same emergency treatment in EU member states as residents. Patients from other EU member states are entitled to claim the same benefits as domestic patients. In 1998, patients' rights in cross-border care were considerably extended by the European Court of Justice. In its Kohll and Decker decisions it ruled that healthcare services provided in another EU member state on a remunerated basis constitute services as defined in the EU Treaty. Measures rendering the reimbursement of amounts paid contingent upon prior consent are obstacles to the free movement of services which need to be justified. These and later rulings were fundamental in that they placed limits on the provisions of the national EU healthcare systems which up to then were exclusively framed by the EU member states.

Germany was one of the first EU member states to implement the decisions of the European Court of Justice on patient mobility into national law by passing the

---

[2] See Chambers et al. (2010, p. 3).

[3] Regulation (EEC) No 1408/71 of the Council of 14 June 1971 on the application of social security schemes to employed persons and their families moving within the Community.

Statutory Health Insurance Modernization Act of 2004.[4] It explicitly allows SHI members to claim inpatient and outpatient treatment in all EU member states and in the additional four member states of the European Free Trade Association (EFTA), i.e. Iceland, Liechtenstein, Norway and Switzerland.[5] Furthermore, the legislation allows sickness funds to directly enter into contracts with healthcare providers in the European Economic Area (EEA) and Switzerland.[6]

The following principles on EU cross-border care are legally implemented[7]:

- All SHI members can claim outpatient treatment to which they would be entitled in Germany in any other EU member state without prior permission.
- All SHI members can claim inpatient treatment to which they would be entitled in Germany in any other EU member state subject to prior permission. This permission must be granted if the German healthcare system cannot provide the treatment within the medically necessary period.
- In both cases, the costs must generally be reimbursed up to the amount which would be reimbursed in Germany if the domestic preconditions for treatment are fulfilled.

These principles are in full accordance with the EU directive 2011/24/EU on the application of patients' rights in cross-border healthcare. The only change in German law necessary to meet this directive was the legal establishment of a national contact point for cross-border healthcare.

The EU patients' rights directive complements the existing regulation (EC) No 883/2004[8] on the coordination of social security systems. This regulation replaced the Council Regulation (EEC) No. 1408/71 and is particularly relevant for frontier workers who are members of the EU healthcare system in the country in which they work but who reside in another EU country. It specifies the access to healthcare in both states.

The directive differs from regulation (EC) No 883/2004 in three aspects: only planned cross-border care treatments are affected; there is no need for an approval for outpatient care prior to the planned treatment; costs of the planned EU cross-border treatment are only borne up to the amount a comparable treatment would have cost in the home country less a deduction for administration.

According to the regulation (EC) No 883/2004, by contrast, cross-border patients obtain healthcare as if they would be insured in the country they visit. If this coverage is more generous than in Germany, sickness funds must therefore

[4] Gesetz zur Modernisierung der gesetzlichen Krankenversicherung dated 14 November 2003 | Bundesgesetzblatt 2003 Part 1 No. 55.

[5] In the following "EU member States" or "EU cross-border care" also includes these European countries.

[6] Social Code Book V, Sect. 140e.

[7] Social Code Book V, Sect. 13 para. 4–6.

[8] Regulation (EC) No 883/2004 of the European Parliament and of the Council on the coordination of social security systems of 29 April 2004.

reimburse it. On the other hand, if copayments or other out-of-pocket-payments apply to patients in that country, the SHI member would have to pay them as well. The German SHI members are allowed to choose between reimbursement according to the EU patients' rights directive and the regulation No. 1408/71 which improves the situation for SHI members for planned cross-border care treatments.[9]

2.3 Sickness Funds and Cross-Border Healthcare

European legislation on cross-border healthcare implies for sickness funds that they must reimburse treatments in other member states. In case of inpatient treatment, sickness funds must also authorize treatment. This can involve considerable administrative costs.[10] To cover these costs, funds are allowed to deduct an administrative surcharge when they reimburse treatment. German legislation also gives funds a potential active role in cross-border healthcare. They are allowed to directly enter into contracts with healthcare providers in the European Economic Area (EEA) and Switzerland.

The currently second largest German statutory health insurance fund, Techniker Krankenkasse (TK) with approximately 8.3 million members in 2013 has taken the most pro-active strategy towards cross-border healthcare. Following the Kohll and Decker rulings of the European Court of Justice, TK started to offer a range of individual services regarding EU cross-border care. TK members who plan to receive medical care in another EU member state have the possibility to be consulted concerning quality, costs, billing, entitlement and SHI-benefits. The consulting on EU cross-border care at TK predominantly takes place in customer consulting offices or by telephone and email through the call and specialist service centers.[11] TK offers a free travel hotline to support its members before and during their stay abroad. Consultation is offered by doctors who are trained in travel and tropical medicine for all questions on cross-border care, both planned and unplanned.

What distinguishes TK from competing sickness funds is its strategy to contract directly with qualified service providers in other EU member states.[12] TK monitors quality of contractual partners and requires, for example, care by physicians and nurses who speak German fluently. Billing directly takes place between the contractual partners and TK, which is to benefit of both SHI members, who do not

---

[9] See Federal Ministry of Health in Germany (2013).

[10] For cross-border care in the Meuse-Rhine eurogio based on regulation No. 1408/71, the additional administrative cost was estimated to be 5 % (Grunwald and Smit 1999).

[11] See Wagner et al. (2011), p. 14.

[12] To some extent, other sickness funds have also contracted with foreign providers. See Nebling and Schemken (2006) for further examples.

have to pay cash in advance, and TK, which does not have to check bills from unknown providers. TK is also able to negotiate prices in exchange for giving the contractual partners better access to its members.

Currently, the TK European Service for Clinics comprises around 135 clinics in eight member states (Austria, Belgium, Czech Republic, Italy, Netherlands, Poland, Portugal and Spain). This service was originally initiated in co-operation with another sickness fund[13] in regions in which a large number of unplanned acute and emergency EU cross-border treatments of TK members occurred like ski injuries in Austria. Furthermore, there are other additional single contracts between TK and EU cross-border healthcare service providers. In all TK partner clinics, it is possible to receive planned inpatient and outpatient cross-border care. Moreover, TK has lately entered contracts with four dental clinics in Hungary and Poland. Finally, TK has contracted with nearly forty spa and rehabilitation facilities which are located in six member states, predominantly in Eastern Europe (Austria, Czech Republic, Hungary, Italy, Poland and Slovakia). TK members receive special conditions, e.g. regarding the accommodation. TK reimburses all the contractually declared therapy costs. On top wellness treatments like e.g. additional massages have to be borne out-of-pocket by the members. These additional treatments tend to be less expensive than in Germany, in particular in Eastern European Countries. Furthermore, TK pays a one-time extra payment of 100 Euro for a cure or rehabilitation treatment which lasts for at least two weeks.

TK has carried out surveys on experiences and expectations concerning EU cross-border care among its members, particularly focusing on patient satisfaction with regard to planned treatments. Its key results are presented in the following section.

3 Germany's Experience with EU Cross-Border Healthcare Prior to the Directive

TK has carried out several EU cross-border surveys. In the following, we present important findings of the two latest surveys published in 2010 and 2012.[14] For the first survey, questionnaires were sent to randomly chosen individuals who used cross-border care in 2009. The second survey asked a random sample of those who obtained cross-border care in 2010. For the 2010 survey, the response rate was 31 %, for the 2012 survey it reached 41 %. About 1 % of TK members claimed reimbursement for cross-border care in a year. The 2010 survey reveals that about 26 % of these had planned to use EU cross-border care. For the 2012 survey, the corresponding share is 19 %. We focus on this group in the following.

---

[13] The sickness fund "AOK Rheinland Hamburg" .

[14] See Wagner et al. (2011, 2013).

3.1 Users of Planned EU Cross-Border Healthcare and Destinations

TK members with planned cross-border treatment are comparatively old on average (see Fig. 1). Nearly 80 % were over 60 years, out of which half were aged 70–79 years, and another 8 % were older than 80 years. Accordingly, the majority are retirees (70 %). The remaining members are employees (16 %), housewives and house husbands (10 %), and self-employed persons (4 %). Both genders are equally represented with 50 %. The level of education of planned cross-border care consumers at TK is high with a share of 54 % having passed their A-levels ("Abitur"), out of which 32 % are university graduates and another 3 % have a doctorate in addition. Since most cross-border care users are retired, the income level is comparatively low: 50 % had a gross income of 1,000–2,500 Euros, another 10 % of 750–1,000 Euros.

Living in border regions is not crucial for planned EU cross-border care: only 7 % of the surveyed TK members stated this as a reason. Only 16 % of TK cross-border care consumers live in border regions, i.e. at a distance of less than 30 km. Another 13 % live more than 30 but less than 60 km from the border. The majority of 71 % live more than 60 km from the border.

One of the main survey findings is that a certain degree of continuity and regularity has emerged: 53 % of the TK members surveyed consume planned EU

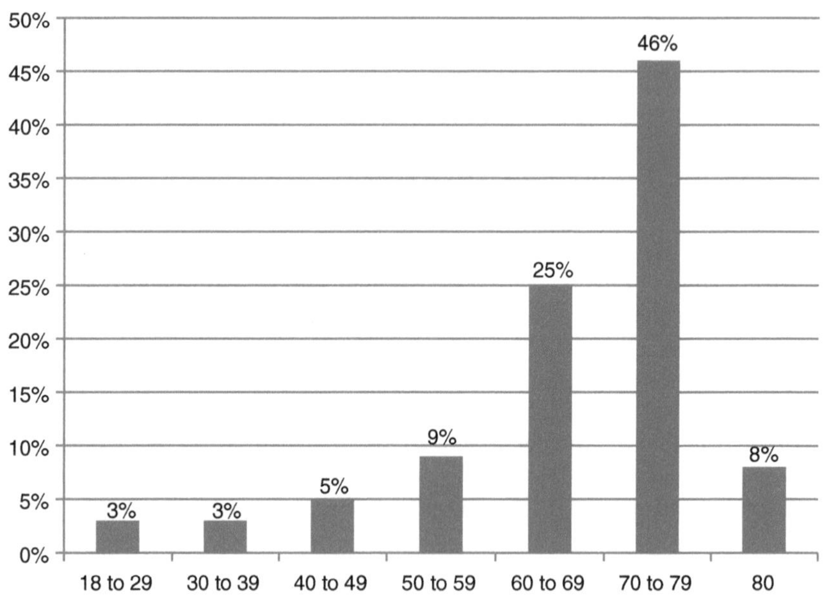

Fig. 1 Age distribution of TK members with planned cross-border treatment. *Source* Wagner et al. (2013)

cross-border care regularly, out of these 39 % regularly do so each year. Another 73 % stated that they get treated deliberately once a year, 12 % twice a year and 16 % three times a year or more.

According to the 2012 survey, Italy is the most popular destination for planned EU cross-border treatment of TK members (16 % in 2010 and 2012). In 2010, this has been Czech Republic (2010: 27 %, 2012: 14 %) and Poland (2010: 20 %, 2012: 15 %). Hungary had a share of 12 % in both surveys. Overall, the Eastern European States including Hungary have slightly lost in popularity. Austria (2010: 6 % to 2012: 10 %) and Switzerland (2010: 5 % to 2012: 8 %), but also Spain (2010: 2 % to 2012: 6 %) and France (2010: 1 % to 2012: 3 %) have gained popularity.

3.2 Treatments Provided

With 51 %, preventive cure and spa treatments were the most frequently demanded types of EU cross-border care. These are, quite uniquely as compared to the other EU healthcare systems, part of the statutory benefits catalog covered by sickness funds in Germany. A trend is the growth in the demand for planned treatments in clinics: the share of members who sought them rose considerably from 4 % in 2010 to 16 % in 2012. Planned dental and orthodontic cross-border treatments ranked third among TK members and grew, too (2010: 7 % and 2012: 11 %). Specialists and general practitioners were noticeably less, but equally often consulted (Fig. 2).

The most frequent medical reasons for TK members to travel abroad for care in 2012 were muscular, bone, and joint diseases (55 %), though the share of members stating this has considerably fallen from a level of around 70 % in 2010. The next most frequent diseases affected the cardiovascular system (11 %), teeth (10 %), respiratory diseases (8 %), cancer (4 %), follow-up treatments of injuries caused by accidents or intoxications (4 %), diabetes (3 %), and skin diseases (3 %).

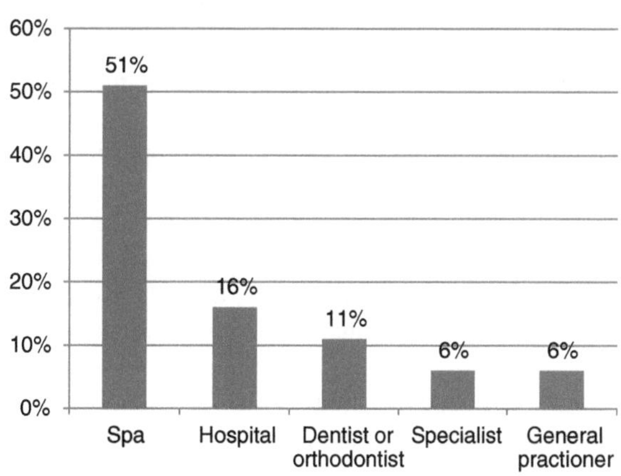

Fig. 2 Treatments by provider, 2012 Survey. *Source* Wagner et al. (2013)

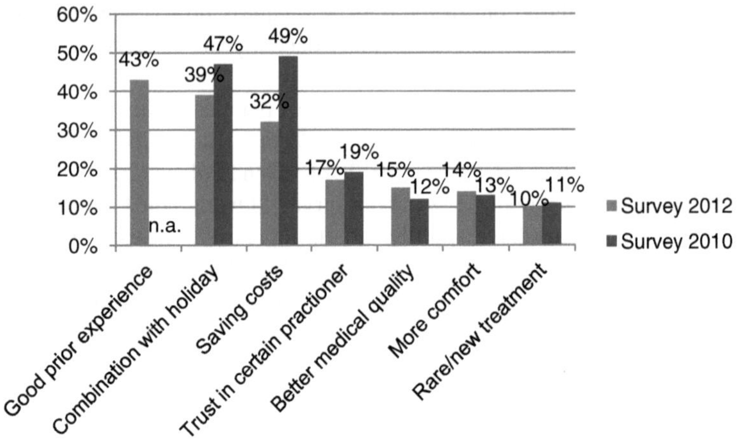

Fig. 3 Motivation for planned cross-border care. Multiple responses possible. *Source* Wagner et al. (2011, 2013)

3.3 Motivation for Planned Cross-Border Treatment

When asked about the motivation to deliberately seek cross-border care, the most common reason in the 2012 survey was good experience with former treatments (43 %, not asked in the 2010 survey). Combining treatment with holiday (2012: 39 %, 2010: 47 %) and cost savings are also important motives (2012: 32 %, 2010: 49 %). A frequently stated reason was also confidence in a cross-border doctor (2012: 17 %, 2010: 19 %). Moreover, in the 2012 survey, 15 % of the surveyed TK members stated to search for better quality and 14 % for better comfort in other EU healthcare systems (Fig. 3).

3.4 Satisfaction with Planned Cross-Border Care

Planned cross-border care among TK members is marked by a high level of satisfaction. This concerns the facilities, the doctors and dentists as well as the therapies and treatment results. The proportion of members who were very satisfied with the facilities amounts to more than 70 % regarding short waiting times for appointment, the cleanliness of rooms, the efficiency of the working processes and the agreeable atmosphere.

The evaluation of the doctors, dentists and quality of the treatment was similarly positive. According to the subjective assessment of the treatment results by the 2012 survey respondents, planned cross-border care in those EU healthcare systems frequently visited by the TK members is of high quality. This is also supported by the fact that less than 2 % stated to have had to undergo a follow-up or further treatment because of unexpected complications or maltreatment. TK members were most satisfied with the medical competence of the doctors (78 %), with the doctors' thoroughness in examination and in treatment (74 %), as well as the comprehensibility of the information given by them (73 %). Up to 85 % were very satisfied with the treatment results from dentists and orthodontists, 79 and 78 % respectively with the treatment results from general practitioners and specialists. Less than 1 % stated throughout the different groups of doctors to be very dissatisfied with the treatment results. Even if some respondents are looking for better treatment and comfort in cross-border care, the high level of satisfaction does not seem to be a consequence of dissatisfaction with German healthcare. 59 % indicated to be rather satisfied with German healthcare, another 15 % were very satisfied. Only 23 % were rather dissatisfied and just 3 % expressed extreme dissatisfaction.

3.5 Summary

The surveys reveal that a small but considerable share of sickness fund members make use of planned cross-border healthcare. The majority of these are retired and well-educated. Spa treatments are most popular but sickness fund members also plan to use hospital, dental and specialist services in other European countries. The three main motivations for obtaining treatment abroad are good prior experience, the possibility to combine the treatment with holidays and saving costs. Sickness fund members state a high satisfaction with planned cross-border care.

4 Possible Effects of Increased EU Cross-Border Healthcare for the German Sickness Fund System

Germany is a country with a fairly high standard of care. Currently, access to care is usually fast. Official waiting lists do not exist except for treatments with a natural shortage such as organ transplantations.[15] However, Germany is also a rapidly aging country. Several studies forecast a large increase in healthcare expenditure per capita, predicting increases in the contribution rate from currently 15.5 % of wage income to far above 20 % (see Postler 2010 for a survey). For example, the study by Breyer and Ulrich (2000), which predicted the current contribution rate quite accurately, states a contribution rate of 23.1 % in 2040. It is

---

[15] Nevertheless, waiting times exist in Germany. Roll et al. (2012) find that SHI patients have to wait on average 16 days to get an appointment with a specialist. Privately insured patients, by contrast, only have to wait 7 days on average. They also found that increased income had a negative effect on waiting time, pointing to an income gradient in the access to health care.

unlikely that this increase in the financial burden will not lead to stronger pressure to keep expenditure under control. Healthcare will probably be more rationed than today, either explicitly or implicitly.[16]

Given this scenario, EU cross-border care presents a challenge to the German sickness fund system in two related respects. First, expenditure control is made more difficult by the possibility of EU cross-border care which de facto introduces fee-for-service remuneration for treatment in foreign countries with little further restrictions. This is in contrast to remuneration in place inside Germany which relies on expenditure control mechanisms. Second, equal access to care may be undermined if those who are able to afford cross-country care can avoid rationing of healthcare in Germany. At the same time, cross-border care can also be a means of saving costs for the German healthcare system. In particular, sickness funds could be given a more active role in organizing and encouraging cross-border care. However, it must be ensured that cross-border care is not used for risk selection by competing sickness funds.

4.1 Cross-Border Care and Expenditure Control

The EU directive of the application of patients' rights and German law require that sickness funds must reimburse up to the amount which they would have paid for the same treatment in Germany. This regulation appears simpler than it is. Even though in Germany fees are specified for treatments, expenditure control measures lead to lower actual payments.

In the inpatient sector, hospitals are mainly paid based on DRGs. The German DRG catalog specifies a relative weight for each DRG.[17] Furthermore, for each federal state a "state basis value" ("Landesbasisfallwert") is specified which is to be multiplied by the relative weight to obtain the DRG payment. On average, about €3.050 are paid per standard case with weight one in 2013. If a person is treated in another country, this would suggest paying the state basis value multiplied by the relative weight of the corresponding DRG in Germany. This, however, would not take into account that there are also agreements on the total volume of services supplied by hospitals in Germany. These are broken down for each hospital in Germany, resulting in target budgets ("Erlösbudgets"). If hospitals exceed these target budgets, payments are considerably reduced. Currently, only 35 % of the DRG payment can then be kept by the hospital. In practice, this is frequently the case. Thus, sickness funds effectively pay less for hospital treatments per case than the payment according to the state basis value. This would

[16] See Zweifel et al. (2009), Sect. 5.5 for a discussion of types of rationing of health care services.

[17] The German DRG system is managed by InEK (Institut für das Entgeltsystem im Krankenhaus). Information can be obtained on its website http://www.g-drg.de.

need to be considered for cross-border care. However, the exact amount paid on average for hospital treatment is known only retrospectively.

For the outpatient sector, the arrangement is even more complicated. For the most part, sickness funds pay capitations to regional physicians' associations. These distribute their total revenue to their members according to a complex catalog which specifies fees for cases as well as for specific services. Once again, these fees are overestimates of actual payments. This is a consequence of the target budgets for physicians ("Regelleistungsvolumina"). If this budget is exceeded, the reimbursement for additional cases is significantly reduced.

At the moment, sickness funds use official fees to calculate the maximum amount for reimbursement in case of cross-border treatment. On average, treatment in foreign countries can therefore be more generously reimbursed than in Germany itself. From the perspective of sickness funds, this is true in a double sense for the outpatient sector. They continue to pay the capitation to the regional physicians' association and pay the cross-border treatment on top. Since the treatment is not provided by the association, sickness funds could in principle claim back the expenditure for cross-border treatment. However, resistance from associations is strong and this has not yet been implemented.

Both in the inpatient and the outpatient sector, cross-border care therefore has the potential to increase overall expenditure of the German sickness fund system because it bypasses the expenditure control mechanisms. Currently, this does not matter too much because overall expenditures for cross-border care are still low. Furthermore, if treatment in other member states is less expensive than in Germany, then the potentially higher reimbursement is not a problem. This holds for most treatments in Eastern European countries but not for countries such as Austria or Switzerland. If cross-border care becomes more widespread with treatment in "expensive" countries, then measures would need to be taken.

4.2 Cross-Border Care and Equal Access to Care

One worry about cross-border care is that it can increase inequality of access to healthcare.[18] In particular, those which are physically and financially able to travel have an advantage over those with complex health conditions or little means.[19] Foreign providers can also have the incentive to attract easily treatable patients by offering spa and wellness treatments on top at a discount. Expensive and immobile patients with difficult and chronic conditions will probably remain in the home country.

This problem will be more severe, the stronger care is rationed in the home country. This holds for both explicit and implicit rationing as long as the treatment

[18] See footnote 16 for evidence for unequal access in Germany.

[19] See Baeten (2011, p. 265).

is part of the benefit package of the public healthcare system.[20] In the case of explicit rationing with waiting lists, European cross-border care opens the possibility of jumping the queue. In particular, the rich and educated have an advantage. With implicit rationing through budgets, similar effects can arise. The EU directive effectively implements fee-for-service reimbursement for cross-border care. As pointed out in the previous section, this undermines standard expenditure control measures. Again, those able to use care in another country are privileged because they can claim fee-for-service remuneration.

The German sickness fund system so far relies mainly on implicit rationing.[21] In the inpatient and outpatient sectors, target budgets reduce the incentives of providers to treat patients by reducing reimbursement once the target budget is reached. With the aging of German society and the forecasted increase in healthcare expenditure per capita, the extent of implicit rationing is likely to increase. This creates the incentives to seek cross-border care and make use of the more favorable fee-for-service remuneration.

4.3 Cross-Border Care, Cost Savings and Risk Selection

The previous sections on expenditure control and access to care emphasized the problems created by cross-border care. However, cross-border care also has beneficial effects. The TK surveys showed that patients are very satisfied with cross-border care. They also tend to obtain care in Eastern European countries where healthcare is less expensive. According to the 2012 survey, 41 % of planned EU cross-border care took place in the Czech Republic, Hungary and Poland. This also saves costs for the German sickness fund system as the reimbursement for care in these countries is lower than for the same treatments in Germany.

These cost savings are more likely to arise if sickness funds take an active role in organizing cross-border care. First, sickness funds can assess the quality of foreign providers and thereby avoid costly follow-up treatments in Germany. Second, they can make sure that providers do not overcharge. This holds for the individual patients, who may face high copayments, and for sickness funds, which want to avoid that foreign providers try to extort the maximum payment from them. In particular, sickness funds can negotiate prices with foreign providers.

This approach has been taken by Techniker Krankenkasse. However, this has been observed with some scrutiny by other players in the German sickness fund system. As anywhere, domestic suppliers are not happy about competition from

[20] Explicit rationing may also exclude a treatment from the benefit package of the public health care system. In this case, however, cross-border care is not reimbursed either. Nevertheless, those with more financial means have an advantage because they can obtain this treatment either at home or abroad.

[21] Breyer (2013) argues that it would be desirable to limit implicit rationing and to move towards explicit rationing.

foreign suppliers who can produce at lower cost. Therefore, TK has been cautious in its cross-border operations and does not actively promote cross-border care but only offers assistance to those who seek it.

Cross-border care also has to be seen in the context of competition among sickness funds. Supporting cross-border care is a promising product differentiation strategy for sickness funds. In principle, this is to the benefit of individuals, who have more choice, and of sickness funds, which can save costs. A potential problem, however, is that promoting cross-border care may be a means of risk selection.[22] In particular, it can be a way to attract comparatively good risks. As the surveys have shown, users of cross-border care tend to be well-educated and in rather good health, a finding corroborated by the fact that they are still mobile. These characteristics are not directly considered in the German risk adjustment formula which uses information on age, gender, reduced earning capacity and morbidity. However, there is an easy way to avoid that cross-border care just pays off because cross-border care users tend to be healthier. The fact that a person uses planned cross-border care could be considered in the risk adjustment formula. If these individuals are less expensive compared to otherwise similar individuals, then the risk-adjusted payment for them should be reduced.

5 Policy Implications

The EU patients' rights directive not only creates new opportunities for patients but also raises problems for the German sickness fund system. The reason is that the directive de facto stipulates fee-for-service reimbursement for cross-border treatment while inside Germany reimbursement is limited by target budgets. Official fees are therefore overestimates of actual payments. This implies that the current system which uses official fees for cross-border care is potentially too generous. If cross-border care becomes more widespread, then sickness funds can face considerable expenditure increases. The fee-for-service reimbursement for cross-border care is also problematic for equity reasons. It puts those at an advantage who are able to seek cross-border care because they are still mobile and possess the means for traveling.

To avoid the adverse effects on expenditure and equity, an important step is to correct the official fees for treatment in Germany for the effects of expenditure control measures. A useful approach is to calculate what is actually paid for treatment in Germany considering that some treatments are reimbursed only partially because they have been provided when the target budget was already exhausted. Since this is known only retrospectively, data from previous years can be used to calculate a reduction due to budgeting restrictions. A problem of this

[22] This would be an instance of indirect risk selection where insurers use the design of the benefit package to attract favorable risks (see Zweifel et al. 2009, Chap. 7).

approach is that it may be regarded as a way of limiting patient mobility in the EU. However, it is in line with the requirement that only the amount must be reimbursed that is paid in the home country.

Cross-border care also requires adjustments within the German sickness fund system. One problem is that sickness funds now pay twice for a patient who seeks outpatient care abroad because they continue to pay the capitation to the regional physicians' association. Payments for cross-border care should therefore be deducted from the associations' budget. At the moment, however, this is difficult because the fees paid for cross-border care can be larger than the fees received by a physician when treating the patient at home. Again, this calls for a recalculation of the remuneration of cross-border care.

Furthermore, attention should be given to the potential of cross-border care as a means for risk selection. A simple approach is to include the fact that a person used planned cross-border care in the risk adjustment scheme. With this policy in place, sickness funds can be given an active role in organizing cross-border care. This is a promising way to ensure that the potential benefits of the EU patients' rights directive are not only to the advantage of the individual patient but also for the sickness fund system as a whole.

Overall, however, a fundamental problem remains. The coexistence of one sector of care which relies on expenditure control mechanisms and implicit rationing (domestic treatment) and an alternative sector of care which is reimbursed fee-for-service (cross-border treatment) is conflictive. If such a choice would exist within a country, the expenditure-controlled sector would likely suffer at the expensive of the fee-for-service sector. Fee-for-service, however, raises its own problems. In particular, it is vulnerable to both moral hazard by patients and supplier-induced demand by physicians. To counter these effects, copayments may have to be increased. It remains to be seen whether cross-border care will create such problems. Up to now, its impact has been limited.

6 Conclusions

Germany provides an interesting case study for possible impacts of the EU directive on the application of patients' rights in cross-border healthcare for outward patient mobility. Already in 2004, most of the directive's requirements were legally implemented in Germany. Since then, sickness fund members can claim treatment to which they would be entitled in Germany in any other EU member state and in the four member states of the European Free Trade Association. Only for inpatient treatment, prior permission is required. Costs must generally be reimbursed up to the amount which would be reimbursed in Germany.

Results from surveys by the German sickness fund with the most pro-active strategy towards cross-border care show that a small but considerable share of sickness fund members makes use of planned cross-border healthcare. The majority of these are retired and well-educated. Spa treatments are most popular but sickness fund members also use hospital, dental and specialist services in other European countries. They are generally highly satisfied with planned cross-border care.

A problem for the German sickness fund system is that cross-border treatment is effectively reimbursed on a fee-for-service basis. By contrast, remuneration inside Germany is subject to expenditure control mechanisms. In particular, official rates used for cross-border care can be too high at the moment since they do not take into account reduction of fees due to targeted budgets. The discrepancy of rationing inside Germany and fee-for-service outside Germany can also be problematic for equity reasons. Those who are able to travel to foreign countries may get access to better treatment because it is more generously reimbursed. To avoid these adverse effects on expenditure and equity, the official fees for treatment in Germany should be corrected for the effects of expenditure control measures.

The problem of higher expenditure does not arise if sickness fund members obtain treatment in less expensive countries, in particular in Eastern Europe. Cost savings can be expected if sickness funds are given an active role in organizing these treatments and in contracting with local providers. To avoid that sickness funds use cross-border care to select healthier patients, the fact that a person used planned cross-border care should be included in the German risk adjustment scheme.

Overall, the German experience with cross-border care shows that patients make use of the possibility of being treated in other European countries and are usually satisfied. It also demonstrates that it is difficult to define the amount which would be reimbursed in Germany for the same treatment. On a fundamental level, there is a tension between resource allocation within Germany and the fee-for-service reimbursement for cross-border care. An open question is the effect of increased cross-border care on the healthcare systems of the countries in which German sickness fund members obtain treatment. On the positive side is the additional income created by patients from Germany. However, patients in these countries may have to wait longer than German patients, in particular in Eastern Europe. This has to be balanced against the incentives created by cross-border care for physicians to stay in their home country rather than to migrate to countries with higher income opportunities.

References

Baeten, R. (2011). Past impacts of cross-border health care. In M. Wismar, W. Palm, J. Figueras, K. Ernst & E. van Ginneken (Eds.), *Cross-border health care in the European Union: Mapping and analysing practices and policies* (pp. 255–287). Observatory study series No. 22, European Observatory on Health Systems and Policies.

Breyer, F., & Ulrich, V. (2000). Gesundheitsausgaben, Alter und medizinischer Fortschritt: Eine Regressionsanalyse. *Jahrbücher für Nationalökonomie und Statistik, 220*(1), 1–17.

Breyer, F. (2013). Implicit versus explicit rationing of health services. *CESifo DICE Report, 11*(1), 7–15.

Chambers, N., Alakeson, V., Lewis, G., Porter, A., & Shaw, S. (2010). *The Techniker Krankenkasse experience: Lessons for commissioners from a successful German statutory health insurer*. The Nuffield Trust for Research and Policy Studies in Health Services.

Federal Ministry of Health in Germany (2013). ("Bundesministerium für Gesundheit"). http://www.bmg.bund.de/glossarbegriffe/p-q/patientenmobilitaetsrichtlinie.html

Grunwald, C., & Smit, R. (1999). *Grensoverschrijdende zorg Zorg op Maat in de Euregio Maas-Rijn; evaluatie van een experiment*. Utrecht: National Hospital Institute.

Nebling, T., & Schemken, H. W. (2006). Cross-border contracting: The German experience. In M. Rosenmöller, M. McKee, & R. Baeten (Eds.), *Patient mobility in the European Union: Learning from experience* (pp. 137–156). Copenhagen: WHO Regional Office for Europe.

Postler, A. (2010). *Nachhaltige Finanzierung der Gesetzlichen Krankenversicherung*. Berlin: Duncker & Humblot.

Roll, K., Stargardt, T., & Schreyögg, J. (2012). The effect of type of insurance and income on waiting time in outpatient care. *Geneva Papers of Risk and Insurance, 37*(4), 609–632.

Wagner, C., Dobrick, K., & Verheyen, F. (2011). EU Cross-border Health Care Survey 2010: Patient Satisfaction, Quality, Information, and Potential. WINEG Wissen 02 Hamburg: Techniker Krankenkasse. http://www.wineg.de.

Wagner, C., Moser, F., Hohn, A., Dobrick, K., & Verheyen, F. (2013). EU Cross-Border Health Care Survey 2012. Planned Cross-Border Care - Physicians and Dentists from TK Insurants Perspective. WINEG Wissen 03. Hamburg: Techniker Krankenkasse.

Zweifel, P., Breyer, F., & Kifmann, M. (2009). *Health economics* (2nd ed.). Berlin: Springer.

The Possible Effects of Health Professional Mobility on Access to Care for Patients

Irene A. Glinos

Abstract The chapter explains how health professional mobility impacts on the resources and capacity available within a health system, and how this affects service delivery and access. The contrasting experiences of destination countries, which receive foreign inflows of health professionals, and of source countries, which loose workforce due to outflows, are illustrated with country examples. The evidence opens the debate on how EU countries compete for health workforce, what this means for resource-strained, crisis-hit Member States, and whether there is any room for intra-European solidarity. The nexus between patient mobility and health professional mobility is moreover highlighted. This take on free mobility in the EU has received little attention, and while evidence is scarce, it calls for careful analysis when considering the possible effects of free movement on access to care in national health systems. The chapter reformulates the question on 'who wins' and 'who looses' from freedom of movement in the EU to turn our attention away from those *who go abroad for care* and instead focus on those *who stay at home*.

Keywords Health professionals · Mobility · Migration · Health workforce · Service delivery · Access to care

JEL Classification F220 · I110 · J610 · H510

I. A. Glinos (✉)
European Observatory on Health Systems and Policies, Rue de l'Autonomie 4 1070 Brussels, Belgium
e-mail: igl@obs.euro.who.int

R. Levaggi and M. Montefiori (eds.), *Health Care Provision and Patient Mobility*, Developments in Health Economics and Public Policy 12,
DOI: 10.1007/978-88-470-5480-6_4, © Springer-Verlag Italia 2014

1 Introduction: Why a Chapter on Health Professional Mobility?

The topic of patient mobility is receiving considerable and growing attention in Europe and elsewhere. As a result of advancements in medicine, transportation, and communication, new options have opened up to patients, and providers around the world are eager to respond to their needs. In the EU, 15 years of supranational jurisprudence and ensuing legislation have put patient mobility firmly on the policy agenda (Palm and Glinos 2010). One intensely debated issue is to what extent public, statutory health systems should pay for planned, non-emergency care obtained in another EU Member State.[1] Advocates of free patient mobility essentially argue that European patients should enjoy the same rights independently of where they live—if providers in another EU country are able to give the 'best' treatment, it is fair that patients have access to it within a union built on freedom of movement. Campaigns such as 'Europe for patients', launched by the European Commission in 2008,[2] capture this logic.

Arguments in favor of pan-European patient mobility and Europe-wide access to care may be emotively strong however they only tell part of the story. Planned health care received in another country is the tip of the health care consumption iceberg. The bulk of health care is consumed and delivered within the borders of national health systems. Out of 28 European countries, only three (Luxembourg, Cyprus and Iceland) spent more than 1 % of health expenditure on services delivered abroad in 2010 (OECD 2012). In a country like the Netherlands, 'imported' care represented 0.8 % of the €36 billion health care budget in 2010.[3] These percentages may well include costs from travelers who need emergency care while abroad. The real question therefore is what happens to the 99 % of patients who receive care within their home system—in what ways does freedom of movement affect access options and care delivery within health systems in the EU?

Patients' access to the treatment they need depends not only on their entitlements (the focus of patient mobility debates) but on whether the appropriate, qualified health professionals are available to deliver the services. A skilled and sufficient health workforce is the backbone of any health system. The migration of health professionals—as opposed to that of patients—is potentially a much more important issue if the objective pursued is to ensure access to care (Glinos 2012).

The chapter will explain how health professional mobility impacts on the resources and capacity available within a health system, and how this affects service delivery and access. The contrasting experiences of destination countries, which receive foreign inflows of health professionals, and of source countries,

---

[1] A distinction is made here between planned care for which the patients travels deliberately to another country, and emergency care which the patient needs while traveling abroad. In this chapter, 'patient mobility' refers only to traveling abroad for planned care.

[2] http://ec.europa.eu/health-eu/europe_for_patients/about/index_en.htm.

[3] Dutch Health Insurance Board. www.cvz.nl/zorgcijfers/zvw-lasten/zvw-lasten.html.

which loose workforce due to outflows, will be illustrated with country examples. The evidence will serve to open the debate on how EU countries compete for health workforce, what this means for resource-strained, crisis-hit Member States, and whether there is any room for intra-European solidarity. This take on free mobility in the EU has received little attention, and while evidence is scarce, it calls for careful analysis when considering the possible effects of free movement on access to care in national health systems. The chapter reformulates the question on 'who wins' and 'who looses' from free mobility in the EU to turn our attention away from those *who go abroad for care* and instead focus on those *who stay at home*.

2 Methodology

The majority of information provided in this chapter derives from the volume 'Health professional mobility and health systems: Evidence from 17 European countries' (2011), edited by Wismar, Maier, Glinos, Dussault and Figueras. The volume covers quantitative and qualitative data on health professional mobility from 17 European countries,[4] as well as quantitative data from an additional eight OECD countries[5] (2008 data or latest year available). The data has been complemented with other scientific sources. Grey literature has been used given the scarcity of official information and the broad media coverage of the topic in particular related to the effects of the economic crisis.

3 The Link Between Health Professional Mobility and Service Delivery

One of the key ways health professional mobility impacts on the performance of health systems is by changing the composition of the health workforce in both sending and receiving countries. These gains and losses may strengthen or weaken the performance of health systems and, while the numbers of health professionals leaving or arriving may appear negligible, produce visible impacts when numbers increase or when mobility continues over years. Health professional mobility also affects a health system's skill-mix, i.e. the range of skills and competences of the total health workforce, since skills travel with the mobile health professional. When these skills are rare and essential, outflows of even small numbers of health professionals can impact on service delivery and access. Health professional

4 Austria, Belgium, Estonia, Finland, France, Germany, Hungary, Italy, Lithuania, Poland, Romania, Serbia, Slovakia, Slovenia, Spain, Turkey, and the UK.

5 Australia, Canada, Ireland, New Zealand, Norway, Sweden, Switzerland, and the USA.

mobility can also affect the geographical distribution of health workers in a country. A disproportionately high outflow from a region may cause or aggravate maldistribution, resulting in under-supplied and underserved areas in which the local population is left without sufficient health professionals. Two factors however limit our understanding of the effects of health professional mobility: the scarcity of accurate, comparable and up-to-date data on in- and outflows to and from countries (in particular outflows); and the indirect impact of health professional mobility on service delivery and health system performance where a complex chain of causalities makes it difficult to ascertain what effects are due to what causes (Wismar et al. 2011).

In the following two sections, evidence from destination and source countries respectively will illustrate the importance of health workforce mobility for access to health services.

4 The Perspective of Destination Countries

The contribution of health professional mobility to service delivery and access to care in destination countries depends on a series of factors:

- the relative weight of the foreign health professionals compared to the total health workforce;
- the specialized skills foreign health professionals bring with them to the destination system;
- the geographical location where foreign health professionals settle down;
- the role and duties foreign health professionals take on;
- the specific population needs which the foreign health professionals respond to.

While the factors are not clear-cut and do overlap to some extent, they are helpful to understand the different ways in which health professional mobility can match the need for extra capacity in the destination country. The dimensions will be looked below.

In countries where foreign health professionals make up a considerable share of the total health workforce they visibly contribute to maintaining service levels. This is the case in Belgium, Portugal, Spain, Austria, Norway, Canada and Sweden where foreign medical doctors represented 10–20 % of the workforce. In four countries (Switzerland, Slovenia, Ireland and the UK) this share reached between 22 and 37 % (Maier et al. 2011). Just as importantly, some countries rely on foreign health care professionals to replenish their workforce. Out of 17 European countries, the UK (42 %) and Belgium (25 %) saw the highest proportions of foreign inflows to the medical workforce in 2008 (ibid). Spain recognized 40 % more foreign degrees (5,383) in medicine than Spanish universities produced (3,841) (López-Valcárcel et al. 2011). In Finland, 43 % of newly licensed dentists in 2006–2008 were foreign-trained, with similar proportions in Austria. One in three nurses entering the nursing workforce in Italy (2008) and one in five in Spain (2007) were foreign-trained or foreign-nationals (Maier et al. 2011). In other health systems, foreign health professionals are particularly numerous in certain specialties. In France, foreign medical doctors make up 3.5 % of the medical workforce but the proportion rises to 9.5 % for anesthetists for whom there is high demand (Delamaire and Schweyer 2011). These data clearly show the importance of health professional mobility for certain health systems; service delivery appears to be dependent on the workforce coming from abroad.

In many health systems, foreign health professionals fill not (only) a general need for increased capacity but (also) fill vacancies in under-served and under-supplied areas. Geographical maldistribution of health workforce is a problem across Europe, leading to the phenomenon called 'medical deserts' (Wismar et al. 2011; Delamaire and Schweyer 2011). Health professional mobility can provide a solution when foreign health professionals settle down in these areas. Germany is a notable example where inflows mainly from eastern European Member States to the sparsely populated and less affluent federal states in the former German Democratic Republic have tripled the number of foreign medical doctors between 2000 and 2008. In particular the German hospital sector has become dependent on the services provided by foreign medical doctors (Ognyanova and Busse 2011). In a peripheral region of the southern Netherlands, Maastricht university hospital employs some 1,050 nurses of which 40 % come from neighboring Belgium. Belgian nurses help alleviate recruitment problems in a region perceived as less attractive than other parts of the Netherlands but is within easy reach from Belgium.[6] In north-eastern Estonia, the small number of doctors and dentists arriving from Russia and Ukraine provide services to the Russian-speaking population and help reduce workforce shortages (Saar and Habicht 2011). In the UK, during the 2003 GP workforce crisis, unfilled posts remained vacant for months and existing GPs had to provide care to 1.5 million extra patients (Blitz 2005). In southeast London, facing a 10 % vacancy rate, 89 general practitioners were recruited mainly from France and Spain to fill some of the 103 open posts (Ballard et al. 2004). In France, non-EU medical doctors are instrumental in maintaining service levels and quality in underserved zones, mainly small cities and poor areas with socio-economic problems (Delamaire and Schweyer 2011).

Foreign health professionals taking up work which is less favored by the domestic workforce and/or which responds to an unmet need of the population is also visible in the tasks and duties performed. In France, many non-EU medical doctors are recruited as associate practitioners (so-called 'PAC' who work under the supervision of a senior doctor) and often take on heavier duties such as night shifts (ibid). Elsewhere, it is the need for elderly and home care services which foreign health workers respond to. In Italy, estimates suggest that between 500,000 to a million undocumented foreign care workers from countries including

[6] Interviews with representative from nursing staff association (28 March 2012) and HR advisor (3 April 2012), Maastricht University Hospital.

Moldova, Ukraine, Romania and Peru look after the nation's aging population, particularly in the wealthier northern regions (Bertinato et al. 2011). In Germany (est. 100,000) and Austria (est. 40,000) undocumented care workers also provide services in elderly people's private homes and offer an alternative to nursing homes which families often cannot afford (Ognyanova and Busse 2011; Offermanns et al. 2011). In these cases, undocumented foreign health workforce responds to patient needs which are not met by the official health system.

Finally, it also recognized that foreign health professionals bring cultural and linguistic diversity to the health workforce which as a consequence better reflects the make-up of increasingly multi-cultural societies, as e.g. noted in Spain and the UK (López-Valcárcel et al. 2011; Young 2011). On the other hand, it is also recognized that destination systems may need to invest in acclimatising foreign health professionals to ensure their smooth integration into the system (Lupieri 2013).

5 The Perspective of Source Countries

Source countries tell the other side of the story. In the health professional mobility equation, what destination countries win source countries loose. These losses can be measured in terms of the number and share of health professional mobility leaving the country; the associated financial costs related to the education and training of health workforce; and the gaps in service delivery and access caused by emigration. Outflow data are by nature patchy because emigrating health professionals do not have to de-register when leaving a country. Despite important data limitations, figures from source countries clearly show that emigration is a non-negligible phenomenon: 9,000 Romanian doctors requested certificates to move to another EU Member State between 2007 and 2010 (Galan et al. 2011); around 2 % of doctors in Estonia and in Hungary have done so annually since 2004 (Saar and Habicht 2011; Eke et al. 2011). By 2008, 6.5 % of Polish doctors and dentists had received certificates to migrate (Kautsch and Czabanowska 2011). 2,650 Slovak nurses (8 % of nursing workforce) went to work in Austria in 2003–2008. It is estimated that around 100 Slovak health professionals leave Slovakia every month which only exacerbates the shortage of 2,000 medical doctors reported by the Slovak hospital association (Beňušová et al. 2011). For source countries, loosing even 1 or 2 % of the workforce is not marginal; it accumulates over the years and may damage service delivery and access where shortages already exist (Glinos et al. 2011).

One needs to zoom in on the level of specialities and individual hospitals to understand how these trends translate into shortages which hamper access to care. In Romania, one of the largest hospitals in Bucharest reported that ca. 10 % of its nursing staff had been recorded as leaving the country to work abroad, and many other hospitals report similar problems. Moreover, the emigration of medical doctors disproportionally touches the north-east of the country which is the most economically deprived region and where medical coverage is the lowest (Galan et al. 2011). In Lithuania, emigrating specialist doctors are often from disciplines with the highest number of vacancies such as gynecology and surgery (Padaiga et al. 2011). Similarly in Poland, emigration contributes to workforce shortages as more doctors from specialties with shortages apply for certification to work abroad: while on average 5.5 % of Polish practising medical doctors had made the administrative requests by late 2008, 19 % of doctors in anesthetics and intensive care and 13 % in emergency care had done so (Kautsch and Czabanowska 2011). In Hungary it is reported that the specialities with highest emigration intentions—anesthetics, intensive care, and general practice—are those most likely to create major bottlenecks in the delivery of health services. It is also noted that if emigration concerns specialties where total numbers are limited but the services crucial (such as pathology) it can seriously jeopardise access to care in a given locality (Eke et al. 2011). Another consequence of emigration is the larger workload and lower work morale among remaining staff which may negatively affect service delivery and access.

Important outflows of health professionals may put the sustainability of a health system at risk, in particular when combined with other workforce challenges such as shortages and attrition (i.e. health professionals leaving the public system or the health sector to work elsewhere). It should also be borne in mind that due to patchy outflow data, we only have a partial picture of the actual numbers of health professionals which source countries loose (Maier et al. 2011).

6 Increasingly Interdependent Countries

The data from destination and source countries show that health professional mobility is a widespread phenomenon which has direct as well as indirect consequences for patients' access to care. Yet the evidence becomes even more compelling when considered in the general health workforce context of shortages and demographic aging. In Wismar et al. (2011), 16 out of 17 European countries face current and/or forecasted workforce shortages either in all professions or in particular specialties (GPs, specialized nurses, anesthesiologist, pediatrician, psychiatrists, internists and general surgeons), whether nationwide or in particular regions or hospitals. And the trend is global. With the Affordable Care Act, an additional 32 million US citizens will be entitled to medical insurance (Wilson 2012). Projections suggest a shortfall of 130,600 doctors in the USA by 2025 and almost a million nurses by 2020 (Association of American Medical Colleges 2010). China is said to currently lack 5 million nurses (Yun et al. 2010). Almost one in two nurses in the UK is expected to retire within the next 10 years. According to European Commission estimates, the EU will lack 230,000 doctors and 590,000 nurses by 2020 (EC 2012). As populations age, so does the health care workforce reinforcing pressures on demand for and supply of care. In 14 out of 17 countries, aging of the health workforce is considered a challenge for the health system. Several countries report difficulties in filling posts whether due to retirement, attrition, emigration or underproduction of health professionals (Wismar et al. 2011).

In this context of local and international shortages, countries become increasingly interdependent. By drawing on a global pool of health workforce which is finite but increasingly mobile, countries compete for health professionals. Actions and policy measures taken in one country have repercussions in other.

For destination countries, the risk is to compromise self-sufficiency. Systems which face under-production of health professionals, wrong skill-mix, geographical maldistribution of health professionals and/or general health workforce shortages may easily become dependent on retaining foreign workforce or on attracting new foreign inflows into the domestic system. Yet reliance on foreign health workforce is not to be a sustainable solution. With EU and global competition for health workforce growing, systems become susceptible to changes in the direction and volumes of flows. Mobility patterns are hard to predict so to be reliant on them is no safe option (Glinos et al. 2011).

When a country fails to plan for and produce the health workforce it needs, someone else picks up the bill. For destination countries, international recruitment can present huge cost savings. It has been estimated that Ireland spent just under €7,000 per internationally recruited medical doctor; by comparison, Irish medical schools receive a state-subsidy of €50,000 per undergraduate medical student (Humphries et al. 2013; Cullen 2012; Irish Medical Organisation 2006). For source countries, outflows of health professionals constitute a major budget drain and lead to a set of problems: gaps in service provision; disruptions to health workforce planning; and a net financial loss considering what it costs to educate and train health professionals. Slovakia spends €59,000 to educate and train a specialist doctor (Beňušová et al. 2011). It is estimated that the losses for Serbia and Montenegro due to medical specialists leaving the country is between US$ 9 and 12 billion (Jekic et al. 2011). Emigration is particularly costly for the country of origin when young health professionals migrate, as noted in Estonia and Hungary, because it hampers the replenishment of the workforce and the home country's return on investments.

7 Crisis and Competition: Is There a Role for Solidarity?

How do health care workforce issues and professional mobility resonate in a context of economic austerity? According to recent data, many EU countries have reduced national health care spending (Mladovsky et al. 2012). In Greece, the 2011 health budget was cut by 1.4 billion EUR, with hospitals and salaries being the hardest hit (Kaitelidou and Kouli 2012). The Irish health sector witnessed a similar cut over the 2 year period 2010–2011 (Thomas and Burke 2012). Romania reduced public health care sector salaries by 25 % and froze all new public sector recruitments in 2010 (Galan et al. 2011). France, Ireland and Lithuania have also reduced health care professionals' salaries. Data (2009 or 2010) from Romania, Hungary and Estonia show an increase in numbers of doctors and nurses leaving (Maier et al. 2011), and anecdotal evidence suggests an increase in Greek medical doctors emigrating e.g. to Germany.

If health professionals move to where salary levels and working conditions are best, how do poorer countries compete? At the international level, Member States of the World Health Organization signed in May 2011 a code on the ethical international recruitment of health professionals. Symbolically this was a huge achievement, but within the EU the logic of free mobility prevails. This leads to a situation where it may be 'unethical' for a destination country such as the UK to recruit from say India but not from Romania. A nurse in the UK can expect to earn around €2,500 per month while the average monthly income of nurses in Romania in 2007 was ca. €364 (Galan et al. 2011; Lupieri 2013). With up to tenfold salary differentials between Member States, it can be a huge challenge for poorer countries to compete. Existing salary differences may widen further—between Member States and between the public system and commercial sector within countries, encouraging health care professionals to leave the country or the public sector. Competition between countries may develop into a concrete concern in particular for medical specializations where there are widespread shortages. anesthesiologist are e.g. in high demand across Europe (shortages noted in France, Austria, Spain, Hungary and Poland) and general practitioners are scarce in some countries (shortages noted in Finland Slovakia, and parts of England) (Wismar et al. 2011).

This might be contributing to a widening asymmetry in Europe. The research carried out by Wismar et al. (2011) showed that most destination countries are in EU15: whereas all EU countries experience health workforce emigration, in EU15 the outflow of health professionals is to a large extent compensated by inflows. New evidence however suggests that in addition to this east-west asymmetry, mobility flows are changing as a result of the economic downturn. Countries which until recently had been net importers of foreign health professionals—such as Spain and Portugal—have seen outflows overtake inflows as a result of worsening employment and working conditions. In Spain, 948 medical doctors requested certification to work abroad in the first 6 months of 2012, compared to 650 requests in all of 2007 (Dussault and Buchan 2014 forthcoming; Lopez-Valcarcel et al. 2011). In Portugal, the number of nurses requesting similar documentation almost doubled. Between January and October 2012, the Portuguese Nursing Council received 3,202 requests compared to 1,724 for the entire 2011 (Dussault and Buchan 2014 forthcoming). On the other hand, the UK has seen a sudden marked increase in nurses from Portugal and Spain (Buchan et al. 2014). One of the effects of the crisis may be to reinforce the 'east to west' and 'south to north' direction of health professional mobility in the EU. Destination countries thus have two characteristics: they appear to be in the EU15 group, and to be those least hit by the economic crisis.

The crisis context also reinforces *the nexus between patient mobility and health professional mobility*. Two important factors tend to put a limit to patient mobility in the EU: patients' overall preference to be treated in their home system by providers they are familiar with and close to their relatives mean that EU citizens generally are reluctant to travel for health care (Glinos et al. 2012). Second, legislation at national and EU levels determine the very specific circumstances under which patients have the right to receive health care abroad paid for by the home system following prior authorization (Regulation 883/2004)[7] or be reimbursed by the home system after obtaining health care abroad (Directive 2011/24/EU)[8]. This relatively strict legal framework does not generally encourage patient mobility. There is however one exception where patient motivations and legal entitlements combine to create a context prone to patient mobility. Earlier studies show that *availability* of care is one of the key drivers which makes patients seek treatment outside their home country, e.g. in the event of waiting times and waiting lists at home (Glinos et al. 2010). The Directive makes clear that EU patients are entitled to receive prior authorization from their home system "when [the specific] health care cannot be provided on its territory within a time limit which is medically justifiable" (Art. 8(5)). This means that countries which experience significant delays in the delivery of health services are at higher risk of seeing outflows of patients seeking treatment abroad. Even more importantly, however, is that *health professional mobility may trigger patient mobility*: if the emigration of health workforce leads to declining service delivery and to delays (e.g. in elective surgery), patients may use their EU entitlements to seek those treatments in another EU Member State with the home country paying for it. This places crisis-hit countries in a particularly vulnerable position—at the risk of health professionals, patients and health care funding leaving the system.

The question then is what consequences cross-border mobility may have for solidarity between EU countries. The stakes are high for both destination countries, which benefit from health professional mobility, and source countries, which bear the costs. While hypothetical at this moment, it seems that the current context of widening asymmetries, workforce shortages and aging has the potential of leading to a hierarchy of European systems where poorer systems face a vicious circle of budget cuts, health professionals leaving the country, and patients experiencing access constraints—whether in the form of lower service levels, delays, higher out-of-pocket expenditure, or being forced to seek health care abroad. With the effects of the economic downturn expected to last for some years to come, one way to address intra-EU solidarity would be for EU Member States to acknowledge the multiple disadvantages which crisis-hit source countries face and set up compensation mechanisms to alleviate their burden. Such measures would

---

[7] Regulation (EC) No 883/2004 of the European Parliament and of the Council of 29 April 2004 on the coordination of social security systems, OJ 23.04. 2004, L166/1 (formerly Regulation EEC No 1408/71).

[8] Directive 2011/24/EU of the European Parliament and of the Council of 9 March 2011 on the application of patients' rights in cross-border health care. Official Journal of the European Union 2011; 54; 45–65. http://eurlex.europa.eu/JOHtml.do?uri=OJ:L:2011:088:SOM:EN:HTML

be in line with the spirit of the WHO code on ethical international recruitment, but be politically contentious and difficult to negotiate at EU level.

8 Conclusions

The current and forecasted shortages of health professionals, the aging of the health workforce and of the population, and the growing asymmetry between resource-rich and resource-strained EU Member States, and the significant effect health professional mobility can have on service delivery are all good reasons to place the mobility of health professionals firmly on research and policy agendas. The pressures and tensions European countries face will likely intensify as health care professionals increasingly move between EU Member States rather than between the EU and third countries (Glinos et al. 2011).

The notion of EU-wide solidarity is often mentioned when patient mobility advocates argue that all EU patients should be entitled to the same quality and range of health services. Yet the evidence presented in this chapter makes a clear case that if we worry about all EU citizens having access to adequate levels of health care, then a good place to start would be to acknowledge the importance of health professional mobility in providing health systems with sufficient and adequate workforce. In the absence of any compensation mechanism between 'winning' and 'loosing' countries, this would mean for destination countries to give up their reliance on foreign inflows and instead produce a sufficient health care workforce. For source countries this would mean taking measures to encourage their health care workforce to stay in the country (Glinos 2012). For all EU countries, the important role health professional mobility plays in health system performance calls for better monitoring of inflows and outflows to inform policy priorities and processes. As the vast majority of Europeans receive treatment in their home health system, the mobility of health care professionals may lead to much larger problems with access to health services than ever envisaged in patient mobility debates.

References

Association of American Medical Colleges. (2010). *The impact of health care reform on the future supply and demand for physicians updated projections through 2025*. Washington: AAMC Centre for Workforce Studies.

Ballard, K. D., Robinson, S. I., & Laurence, P. B. (2004). Why do general practitioners from France choose to work in London practices? A qualitative study. *British Journal of General Practice,54*(Oct), 747–752.

Beňušová, K., et al. (2011). Regaining self-sufficiency: Slovakia and the challenges of health professionals leaving the country. In M. Wismar, C. B. Maier, I. A. Glinos, G. Dussault, & J. Figueras (Eds.), *Health professional mobility and health systems: Evidence from 17 European countries. Observatory Studies Series 23*. Copenhagen: European Observatory on Health Systems and Policies.

Bertinato, L., Glinos, I. A., Boscolo, E., & Ciato, L. (2011). Oversupplying doctors but seeking carers: Italy's demographic challenges and health professional mobility. In M. Wismar et al. (Eds.), *Health professional mobility and health systems: Evidence from 17 European countries. Observatory Studies Series No. 23*. European Observatory on Health Systems and Policies/ World Health Organization.

Blitz, B. K. (2005). 'Brain circulation': The Spanish medical profession and international recruitment in the United Kingdom. *Journal of European Social Policy,15*(4), 363–379.

Buchan, J., May, F., & Dussault, G. (2014). Nursing workforce policy and the economic crisis: A global overview. *Journal of Nursing Scholarship, 45*(3), 298–307. doi:10.1111/jnu.12028, http://onlinelibrary.wiley.com/doi/10.1111/jnu.12028/abstract.

Cullen, P. (2012). Nearly €2m spent on recruiting doctors in India and Pakistan. *The Irish Times*, 16 Feb.

Delamaire, M. L., & Schweyer, F. X. (2011). Nationally moderate, locally significant: France and health professional mobility from far and near. In M. Wismar, C. B. Maier, I. A. Glinos, G. Dussault, & J. Figueras (Eds.), *Health professional mobility and health systems: Evidence from 17 European countries. Observatory Studies Series 23*. Copenhagen: European Observatory on Health Systems and Policies.

Dussault, G., & Buchan, J. (2014 forthcoming). The economic crisis in the EU: impact on health workforce mobility. In J. Buchan, M. Wismar, I. A. Glinos, & J. Bremner (Eds.), *Health professional mobility in a changing Europe: New dynamics, mobile individuals and policy responses (working title). Observatory Studies Series*. Copenhagen: European Observatory on Health Systems and Policies.

EC. (2012). *Commission staff working document on an action plan for the EU health workforce SWD (2012) 93 final*, Strasbourg. Retrieved April 18, 2012 from http://ec.europa.eu/dgs/health_consumer/docs/swd_ap_eu_healthcare_workforce_en.pdf.

Eke, E., Girasek, E., & Szocska, M. (2011). From melting pot to laboratory of change in central Europe: Hungary and health workforce migration. In M. Wismar, C. B. Maier, I. A. Glinos, G. Dussault, & J. Figueras (Eds.), *Health professional mobility and health systems: Evidence from 17 European countries. Observatory Studies Series 23*. Copenhagen: European Observatory on Health Systems and Policies.

Galan, A., Olsavszky, V., & Vladescu, C. (2011). Emergent challenges of health professional emigration: Romania's accession to the EU. In M. Wismar, C. B. Maier, I. A. Glinos, G. Dussault, & J. Figueras (Eds.), *Health professional mobility and health systems: Evidence from 17 European countries. Observatory Studies Series 23*. Copenhagen: European Observatory on Health Systems and Policies.

Glinos, I. A., Baeten, R., Helble, M., & Maarse, H. (2010). A typology of cross-border patient mobility. *Health and Place,16*(6), 1145–1155. 2010 Nov. doi: 10.1016/j.healthplace.2010.08.001.

Glinos, I. A., Wismar, M., Maier, C. B., Palm, W., & Figueras, J. (2011). Health professional mobility and health systems in Europe: conclusions from the case-studies. In M. Wismar et al. (Eds.), *Health professional mobility and health systems: Evidence from 17 European countries. Observatory Studies Series No. 23*. Copenhagen: European Observatory on Health Systems and Policies/ World Health Organization.

Glinos, I. A. (2012). Worrying about the wrong thing: patient mobility versus health professional mobility. *Journal of Health Services Research and Policy, 17*(4), 254–256. doi:10.1258/jhsrp.2012.012018.

Glinos, I. A., Doering, N., & Maarse, H. (2012). traveling home for treatment and EU patients' rights to care abroad: results of a survey among German students at Maastricht University. *Health Policy 2012, 105*(1), 38–45. doi:10.1016/j.healthpol.2011.12.008.

Humphries, N., Tyrrell, E., McAleese, S., Bidwell, P., Thomas, S., Normand, C., & Brugha, R. (2013 forthcoming). *Brain gain, waste and drain-a qualitative study of non-EU migrant doctors in Ireland*. Human Resources for Health.

Irish Medical Organisation. (2006). IMO Position Paper on Medical Schools. 2006, Irish Medical Organisation, Dublin.

Jekic, I. et al. (2011). Geopolitics, economic downturn and oversupply of medical doctors: serbia's emigrating health professionals. In M. Wismar, C. B. Maier, I. A. Glinos, G. Dussault, & J. Figueras (Eds.), *Health professional mobility and health systems: Evidence from 17 European countries. Observatory Studies Series 23*. Copenhagen: European Observatory on Health Systems and Policies.

Kaitelidou, D., & Kouli, E. (2012). Greece: The health system in a time of crisis. *Eurohealth (incorporating Eurohealth Observer),18*(1), 12–14.

Kautsch, M., & Czabanowska, K. (2011). When the grass gets greener at home: Poland's changing incentives for health professional mobility. In M. Wismar, C. B. Maier, I. A. Glinos, G. Dussault, & J. Figueras (Eds.), *Health professional mobility and health systems: Evidence from 17 European countries. Observatory Studies Series 23*. Copenhagen: European Observatory on Health Systems and Policies.

López-Valcárcel, B. G., Barber Perez, P., & Quintana, C. D. D. (2011). Opportunities in an expanding health service: Spain between Latin America and Europe. In M. Wismar, C. B. Maier, I. A. Glinos, G. Dussault, & J. Figueras (Eds.), *Health professional mobility and health systems: Evidence from 17 European countries. Observatory Studies Series 23*. Copenhagen: European Observatory on Health Systems and Policies.

Lupieri, S. (2013). Where have Europe's nurses gone? CNN Retrieved May 21 2013, from http://edition.cnn.com/2013/05/21/health/europe-nursing-shortage/index.html?hpt=hp_c3.

Maier, C. B., Glinos, I. A., Wismar, M., Bremner, J., Dussault, G., & Figueras, J. (2011). Cross-country analysis of health professional mobility in Europe: the results. In Wismar et al. (Eds.), *Health professional mobility and health systems: Evidence from 17 European countries. Observatory Studies Series No. 23*. Copenhagen: European Observatory on Health Systems and Policies/ World Health Organization.

Mladovsky, P., Srivastava, D., Cylus, J., Karanikolos, M., Evetovits, T., Thomson, S., & McKee, M. (2012). Health policy responses to the financial crisis in Europe. *Policy summary 5*. Copenhagen: World Health Organization on behalf of the European Observatory on Health Systems and Policies.

Offermanns, G., Malle, E.M., & Jusic, M. (2011). Mobility, language and neighbours: Austria as source and destination country. In M. Wismar, C. B. Maier, I. A. Glinos, G. Dussault, & J. Figueras (Eds.), *Health professional mobility and health systems: Evidence from 17 European countries. Observatory Studies Series 23*. Copenhagen: European Observatory on Health Systems and Policies.

OECD. (2012). *Health at a glance: Europe 2012*. Paris: OECD Publishing. Retrieved from doi:10.1787/9789264183896-en.

Ognyanova, D., & Busse, R. (2011). A destination and a source: Germany manages regional health workforce dispraties with foreign medical doctors. In M. Wismar, C. B. Maier, I. A. Glinos, G. Dussault, & J. Figueras (Eds.), *Health professional mobility and health systems: Evidence from 17 European countries. Observatory Studies Series 23*. Copenhagen: European Observatory on Health Systems and Policies.

Padaiga, Z., Pukas, M., & Starkiene, L. (2011). Awareness, planning and retention: Lithuania's approach to managing health professional mobility. In M. Wismar et al (Eds.) *Health professional mobility and health systems: Evidence from 17 European countries. Observatory Studies Series No. 23*. European Observatory on Health Systems and Policies/ World Health Organization.

Palm, W., & Glinos, I. A. (2010). Enabling patient mobility in the EU: Between free movement and coordination. In E. Mossialos, G. Permanand, R. Baeten, & T. Hervey (Eds.),*Health systems governance in Europe: The role of EU law and policy* (pp. 509–560). Cambridge: Cambridge University Press.

Saar, P., & Habicht, J. (2011). Migration and attrition: Estonia's health sector and cross-border mobility to its northern neighbour. In M. Wismar, C. B. Maier, I. A. Glinos, G. Dussault, & J. Figueras (Eds.), *Health professional mobility and health systems: Evidence from 17 European countries. Observatory Studies Series 23*. Copenhagen: European Observatory on Health Systems and Policies.

Thomas, S., & Burke, S. (2012). Coping with austerity in the Irish health system. *Eurohealth incorporating Eurohealth Observer,18*(1), 7–9.

Wilson, J. (2012). Your health care is covered, but who's going to treat you? *CNN* Retrieved from June 29, 2012, from http://edition.cnn.com/2012/06/29/health/doctor-shortage-affordable-care-act/index.html.

Wismar, M., Maier, C. B., Glinos, I. A., Dussault, G., & Figueras, J. (Eds.). (2011). *Health professional mobility and health systems: Evidence from 17 European countries. Observatory Studies Series 23*. Copenhagen: European Observatory on Health Systems and Policies.

Yun, H., Jie, S., & Anli, J. (2010). Nursing shortage in China: State, causes, and strategy. *Nursing Outlook,58*, 122–128.

Young, R. (2011). A major destination country: The United Kingdom and its changing recruitment policies. In M. Wismar, C. B. Maier, I. A. Glinos, G. Dussault, & J. Figueras (Eds.), *Health professional mobility and health systems: Evidence from 17 European countries. Observatory Studies Series 23*. Copenhagen: European Observatory on Health Systems and Policies.

Patient Choice and Mobility in the UK Health System: Internal and External Markets

Mark Dusheiko

Abstract The National Health Service (NHS) has been the body of the health care system in the United Kingdom (UK) for over 60 years and has sought to provide the population with a high quality service free of user charges for most services. The information age has seen the NHS rapidly transformed from a socialist, centrally planned and publicly provided system to a more market based system orientated towards patients as consumers. The forces of globalization have provided patients in the UK with greater choice in their health care provision, with NHS treatment now offered from any public or approved private provider and the possibility of treatment anywhere in the European Economic Area (EEA) or possibly further. The financial crisis, a large government deficit and austerity public spending policies have imposed a tight budget constraint on the NHS at a time of increasing demand for health care and population pressure. Hence, further rationing of care could imply that patients are incentivised to seek private treatment outside the constraints of the NHS, where the possibility of much greater choice exists in an increasingly globally competitive health care market. This chapter examines the evidence on the response of patients to the possibilities of increased choice and mobility within the internal NHS and external overseas health care markets. It also considers the relationships between patient mobility, health care provision and health policy. Patients are more mobile and willing to travel further to obtain better care outcomes and value for money, but are exposed to greater risk.

M. Dusheiko (✉)
Centre for Health Economics, University of York, Heslington, York, UK
e-mail: mark.dusheiko@york.ac.uk

M. Dusheiko
Institute of Health Economics and Management, University of Lausanne, Lausanne, Switzerland

R. Levaggi and M. Montefiori (eds.), *Health Care Provision and Patient Mobility*, Developments in Health Economics and Public Policy 12,
DOI: 10.1007/978-88-470-5480-6_5, © Springer-Verlag Italia 2014

Keywords Patient mobility · Patient choice · Medical tourism · Health care markets · Competition · Travel times

JEL Classification H42 · I18 · I21

1 The Modern UK National Health Service (NHS) and Development of Health Care Markets, Patient Choice and Mobility

The United Kingdom (UK) National Health Service (NHS) came into existence on 5th July 1948 and was founded on the principles of providing universal, equitable, comprehensive, high quality care throughout the country free at the point of delivery (Delamothe 2008a). NHS care is centrally financed out of general tax revenue and government borrowing with 80.3 % coming from taxation and 8.4 % from National Insurance. The system is financed by a global health care budget set by fixed government contributions to health care spending as a proportion of GDP. Hence, the amount of spending is a matter of government policy and political choice. The UK has controlled spending more effectively during its first 50 years than any other health care system in the world, currently spending 8.2 % of GDP on health care, representing 20 % of total government spending (Appleby 2013). The UK spent on average $3,433 (£2,247) per capita on health care in 2010 which is little more than the OECD average of $3,268 (£2,140) with spending growing at 5.2 % between 2000 and 2009. However, since 2010 austerity has meant frozen spending growth.

The state owns most health care resources (e.g. hospitals) and employs public sector workers on a salaried basis or centrally contracts independent primary care doctors, currently employing 5 % of the population. A health care needs based resource allocation formula devolves budgets to local health care commissioners based on expected population health care expenditures who then purchase necessary services from approved providers (Dixon et al. 2011). The system has an established network of over 8,000 gatekeeping general practices employing on average 3.5 general practitioners (GPs) that control access to non-urgent, specialist treatment provided primarily in hospitals. The UK has 2.7 practicing physicians and 9.6 nurses per 1000 with 2.4 acute care hospital beds. Compared to the OECD average, the UK has more nurses, but there are fewer doctors and hospital beds. Access to diagnostic imaging equipment in 2010 was less than the OECD average with 5.9 MRI and 8.2 CT scanners per million people compared with 12.5 and 22.6 in the OECD (2012). The UK recruits a high proportion of physicians and nurses from abroad (around 37.5 and 8 %, respectively) (OECD 2010).

Since the inception of the NHS, care has had to be rationed. This has been achieved mainly with non-price rationing with significant waiting lists for non-urgent consultation and treatments. The rationing is soft as private insurance and private health care providers are allowed (Delamothe 2008a). The private health care insurance market is mainly offered as an employment benefit in the private sector. Roughly 16 % of the population has private health insurance (IPSOS/MORI 2012). NHS physicians are allowed to work in the public and private sectors and to treat patients in private hospitals or in private NHS facilities. Patient charges were introduced in 1951 for dentistry, ophthalmic services, and pharmaceutical prescriptions. Long term care is privately financed, but there are means tested benefits.

Service rationing occurs implicitly by gatekeeping GPs, physicians, nurses and hospital managers. More explicit rationing takes place in local health authority organizations responsible for purchasing health care. In order to improve the consistency and transparency of health care rationing and to reduce the extent of geographic inequality in access and availability of NHS treatments, the National Institute for Health and Clinical Excellence (NICE) in England and Wales (in Scotland the Scottish Medical Board) was founded in 1999. It acts as the key agency for considering evidence on clinical and cost-effectiveness in order to recommend new treatments for reimbursement on the NHS based on what the NHS is willing to pay. It also provides explicit, transparent and deliberative guidelines on medicines, treatments and procedures. However, budgetary constraints at the local level, limited funding to all NICE recommendations as well as maintenance of existing services implies local rationing still takes place introducing a 'postcode lottery' in access to health care (Drummond and Sorenson 2009). The rationing of NHS treatment and significant geographic variation in access to treatments, new technologies as well as quality of care may motivate patients to be mobile and to have the possibility of exercising choice in where and how they are treated in the NHS or abroad (Laura et al. 2013).

Whether the package of services the NHS has provided is adequate and the degree to which geographic variation in the financing and provision of the service is acceptable has become a major policy concern, with growing health care consumerism making these questions a high priority. Between 1972 and 1998 Britain spent £220 billion (€280 billion; $430 billion) less on health care than the European Union (EU) average. Following recommendations in the Wanless review in 2002, which called for substantial increases in resources for health and social care, the then Prime Minister Tony Blair promised increases in NHS spending from 6.8 % of gross domestic product in 1997 to 9.2 % in 2008 (Wanless 2002; Delmothe 2008b). The tight control of NHS spending was reflected in the British Social Attitudes survey (Appleby 2012), which indicated that for the period 1983–2000, between 60 and 80 % of respondents placed health as the first or second priority for extra government spending. However, following the increase in NHS spending to over 9 % of GDP by 2009, this had fallen from 83 % in 2000 to 68 % in 2011.

1.1 The Evolution of Patient Choice in the Changing NHS Health Care Economy

At the start of the twentieth century wealthier patients had access to a choice of fee charging generalist or specialist physicians who would make house visits as well as to independent teaching hospitals provided by the voluntary sector, which tended to be concentrated in London and other wealthier commercial or industrial cities. For the working classes and destitute access to adequate medical care was restricted by financial means and, although the Poor Law provided medical care, it was limited and often inadequate with patients admitted to infirmaries or infectious disease hospitals. The compulsory National Insurance act of 1911 vastly improved choice and access for the working and middle classes, but choice was restricted to physicians enrolled in the 'Panel List'. The number and quality of municipal hospitals started to improve with access to everyone for a fee adjusted for ability to pay (Leathard 2000).

The Introduction of the NHS established central planning and ensured that local health care services were more uniformly distributed across the country with services provided for communities. Choice for patients was very limited. Individuals were able to choose their GP, dentist and optician, but choice did not extend any further. Over time the NHS moved away from a highly bureaucratic, consensus driven health care economy dominated by the medical profession to a more competitive market for health care services in which independent NHS Hospital Trusts and private providers competed for care contracts from decentralised purchasers and patients directly.

In 1974, the NHS underwent its first major reorganization, becoming autonomous from local government with control given to GPs, hospitals and community care. One intention was for the service to become more responsive to the needs of its users. Patients started to be seen as consumers in relation to health services and to be given some choice in the treatment they received, including being able to access treatment privately at NHS hospitals or contracted by the NHS at private providers (Newdick and Smith 2010).

Concerns about long waiting times and lack of responsiveness to patient's needs lead to the creation of the Conservative Thatcher government's internal market reforms 'Working for patients' (DHSS 1989), which altered health care purchasing arrangements by separating the provision and payment for services. During the 1990s the provider side of the NHS (mainly hospitals) was gradually re-organized into more independent NHS Trusts while purchasing was devolved to regional Health Authorities (HA) and voluntary GP fundholding practices with budgets to reimburse hospitals on a per case basis with prices allowed to vary across providers. The extent of increased choice and responsiveness to local needs, however, was mainly under the control of the HAs and GP fundholders acting as agents for the patient. Evidence suggests that contracting was based on price and waiting times and not quality of care as quality information was less available and incentives were stronger to contract on price and waiting times (Propper 2012). GP fundholders seemed to be more effective purchasers obtaining lower waiting times for their patients and using more providers (Dusheiko et al. 2004, 2008). There was also an effort to facilitate patient choice between GPs with efforts to provide patients with more information about GP practice characteristics, prevent GPs from hindering switching, and increased capitation payments to encourage GPs to attract patients. The information about quality of GP practices was limited and GPs were able to refuse patients based on full lists, designated catchment areas or other criteria. The extent of patient switching of GP practices was limited (Dixon et al. 1997; Masiero 2002). These were the first instances of patient choice combined with financial incentives being used to improve quality and efficiency (British Medical Association 2012).

Choice was not an immediate priority for Tony Blair's newly elected Labour government. GP fundholding was abolished in 1999 and purchasing devolved to Primary Care Trusts (PCTs) intended to create local health care partnerships. Greater decision making and spending power was also devolved to Scotland, Wales, and Northern Ireland, resulting in different health policies across the UK since the year 1997. However, from 2002 the Blair Government embarked on large scale reform of the system with greater patient choice as a central component (Department of Health 2005). In order to meet excess demand the government looked to encourage greater patient driven consumerism and mobility within national borders among domestic providers, and allow the entry of multinational private sector providers to encourage greater responsiveness to patient demand and to increase competition in the health service. There were some initiatives to take advantage of changes in EU legislation in 2000 and to encourage patient mobility by outsourcing treatment with EU partner providers, which was piloted between 2001 and 2005. Plans were announced to offer patients who were already on waiting lists opportunities to choose alternative providers, the scheme was piloted in London (Dawson 2007), but before the evaluation was complete the policy of patient choice of provider at the point of referral had been established. From January 2006 all patients requiring hospital treatment were able to choose from at least four different NHS providers. To support patient choice and competition it was necessary to have spare capacity and incentives to increase productivity and reduce waiting times, hence the Government introduced powerful market solutions for elective care and diagnostic services with new mechanisms including Payment by Results (PbR), which changed the system of hospital payment introducing a fixed tariff prospective payment for treatment (Department of Health 2002).

NHS Foundation Trusts (FT) were established under the Health and Social Care Act 2003 to allow NHS Trusts more independence from central government control and performance management and financial freedoms such as the ability to retain and invest surplus revenue. Greater autonomy was to enable local decision making, responsiveness, innovation and efficiency with incentives to attract patients to increase revenues. Their governance models included members of the public and elected local 'governors' to scrutinize the work of the executive board. An independent regulator called Monitor was established in 2004 with responsibility for approving the financial and governance arrangements of FT. By 2012 there were 98 FT in the acute sector in England representing around 40 % of all acute trusts. The NHS was allowed to negotiate greater use of the spare capacity in the private sector to meet demand followed in 2004 by tendering of contracts for new Independent Sector Treatment Centres (ISTC). The market focused on elective surgery for mainly orthopedics and ophthalmology. In order to encourage entry of new private providers, payments to ISTC were set around 11 % higher than NHS organizations. First wave ISTC were paid a fixed activity revenue stream for 5 years regardless of whether the activity was undertaken. PCT purchasers and NHS hospitals were forced to contract with ISTC and there was concern that the program was not good value for money. The costs to the Department of Health (DH) of establishing the first and second phases of the ISTC program was £146 million at the end of 2006/07 (Health care Commission 2008).

The electronic appointments system Choose and Book was also introduced allowing first outpatient hospital appointments to be booked online or by telephone. By 2008 patients were offered 'free choice' of any NHS or registered independent sector provider for routine elective care. Lord Darzi's NHS Next Stage Review in the same year aimed to increase the health services responsiveness to patients, giving them more control and influence over their health and health care such as the right to choose treatments and providers as well as information on quality through the NHS Choices website. It envisaged a personalized NHS organized around the needs of individuals with greater local provision of services (Department of Health 2008a, b).

By the time the Coalition Government came to power in 2010, the concept of competition and patient choice in the NHS was firmly established. The White Paper 'Equity and excellence: Liberating the NHS' (2010) and the new Health and Social Care Act (2012) gave patients greater choice and control over their care, including choice of referral to named consultant teams; maternity care; care for long-term conditions and end of life care. Improving information and support for patients has also been identified as a key component of enabling meaningful patient choices. The Government has committed to rolling-out personal health budgets for patients with long term conditions to purchase and manage their health care with the aim that services would better meet their needs and to give patients the right to challenge poor quality services or lack of choice. There will be greater choice of services from "Any Qualified Provider (AQP)" in the public or private sectors with responsibility for purchasing local services delegated to Clinical Commissioning Groups (CCGs) made up of GP networks and supporting organizations responsible for the allocation of £65 billion of health spending from April 2013. By 2014 the majority of NHS Trusts should be FT or a 'franchise' of an FT. Regulation of FT will be undertaken separately by Monitor (economic), The NHS Trust Development Authority (governance) and The Care Quality Commission (CQC) will focus on quality of care through inspection and patient feedback. Ultimately, FT will have to manage their governance and financial performance without oversight. Moreover, they will be placed under a 'duty of candour' to be transparent in admitting mistakes; will hold board meetings in public and will produce separate accounts for their NHS and privately-funded activities.

2 Analysis of Patient Choice and Mobility in the NHS Internal Market

2.1 Attitudes, Adoption and Consequences of Patient Choice

The emergence of the patient choice policy, along with initiatives to improve the availability and use of health care quality and performance information in the NHS provides an opportunity to better understand to what extent patients value greater choice, and how they exercise their choices about where to be treated. The problems pervasive in health care in particular uncertainty about health status and treatment effectiveness, asymmetric information between patients and medical practitioners as well as the agency relationship reduce a patient's ability to exercise their preferences and to make informed choices (Evans 1984).

Patient choice requires patients to be aware of the possibility of choosing their health care provider; to be willing to search for different providers and treatment options; to be able to discuss and decide on a preferred provider; and willing and able to travel further for care, as patient mobility is a necessary factor influencing the effectiveness of patient choice policies. Greater choice may come at a greater cost in terms of transaction costs, search costs, switching costs as well as travel costs. Hence, it is important to investigate whether patients are aware of choice, whether they value choice, whether choice was offered or available, how it was undertaken in terms of the sources of information and relative influences in the decision process. It is necessary to know the extent to which choice changes patient's treatment decisions such as which providers they use, the types of treatments chosen and the frequency to which care is utilized. Finally, to evaluate the efficiency of choice policies evidence is required on how choice affects patient health outcomes and satisfaction with the delivery of health care relative to the additional costs of greater choice.

The choice process within an internal health system's market may be different to that when patients are investigating treatment options abroad, as there is a centralized policy with guidance and support from health care providers and administrators to facilitate mobility. The patchy implementation of the patient choice policy in England, however, as well as varying degrees of information disseminated to patients and clinicians meant that there was scope for patients to exercise their initiative in making decisions about where and how to be treated, which was probably not too dissimilar to patient mobility decisions to visit providers in external health care markets abroad. It is therefore likely that significant insights into our understanding of patient mobility can be gained from analysing the evidence from these recent reforms in the English NHS.

A number of studies evaluating the impacts of patient choice have been undertaken following the implementation of the policy in the UK NHS, which provides some insights on an under-researched area. The 2005 British Social Attitudes survey found that 65 % of people wanted to be able to choose their treatment, 63 % their hospital and 53 % the date and time of their appointment. These pro-choice sentiments, however, were contradicted by the Picker Institute who found little interest in choice among a sample of patients admitted to hospital, relative to other questions about their quality of care (Boyd 2007). The Health care Commission (2008) highlighted the need for high quality and accessible local hospitals and concluded patient choice was having a limited impact. Their field-work suggested there would not be a large amount of patient movement due to choice, but there was a change in attitude among service providers. The impact of choice on the quality of elective care appeared limited because patients had little outcome data on health care providers, implying choice alone would not ensure better quality relative to effective contracting and regulatory processes. The 2012 survey of Public Perceptions of the NHS (IPSOS/MORI 2012) appears to support this conclusion as it found that only 4 % of individuals in the general population survey felt there was not enough choice or say in treatment decisions. Alternatively, it indicates that the choice policies have been successful, although only 6 % of respondents were aware that patient choice policies are one of the main organizational changes to the NHS.

Dixon et al. (2010) interviewed patients, GPs and senior executives from hospitals in four PCTs in England between August 2008 and September 2009. Three quarters of respondents said choice was important with choice valued more highly by the elderly, those with no qualifications and from a non-white background. Below half (45 %) of patients were aware of choice of any hospital, which was higher among the elderly and carers. There were differences between the views of patients and their GPs, who thought patients were unaware of choice and did not consider it important. This was reflected by the fact that only half of patients recalled being offered choice by their GPs. For patients offered choice, over two-thirds (69 %) chose their local provider, and 8 % more patients went to a non-local hospital compared to patients not offered choice. Patients attending a non-local hospital were older and more educated and living in more rural areas. Patients were more likely to choose hospitals that were clean, with good facilities and high standards of care and relied on their own experience or that of friends and family (51 %) as well as their GP (36 %). GPs advised patients based on their experience and knowledge from relationships with specialist consultants, feedback from patients and awareness of systemic problems at particular hospitals, but distrusted statistical information about comparative performance.

The National Patient Choice Survey (Department of Health 2010) evaluated the uptake and influences of patient choice as well as some of the consequences for patients across England referred to specialist care from the initiation of the national choice policy in 2006 until March 2010.

The survey revealed significant inequality in the implementation and availability of patient choice. Figure 1 shows that after 4 years of the patient choice policy just over half (54 %) of patients were aware of the possibility to choose their secondary care provider prior to being referred, and the proportion of individuals who were actually offered a choice or had it discussed by their referring health care provider was 49 % in 2010 compared to 30 % in 2006. The levelling off of the uptake of choice may be due to the reluctance among some general practices or geographic regions to promote choice or for certain patient groups to actively ask for choice. Patients aware of choice were more likely to be offered choice (63 % compared to 32 %). However, as awareness has increased, patients actually offered choice levelled off, implying that GP practices are not fully implementing choice. It may also be explained by the fact that choice was not considered relevant for all types of referral. Dixon et al. (2010) reported some resistance, even among 'enthusiastic' GPs, to offering choice to every patient regardless of circumstances. GPs were more willing to let patients choose for more routine referrals, but were more authoritarian when more specialist treatment was required. Patient choice was also geographically dispersed with 18 % of PCT offering choice to below 30 %, while 52 % of PCT offered choice to at least 50 % of their patients in 2010. White women aged between 34 and 64 and white men aged 55–64 and free from a long-standing illness or disability were slightly more likely to be offered choice than other types of patients.

Patients offered choice were 8 % more likely to access a provider they preferred. A higher proportion of patients not offered choice had no preference for

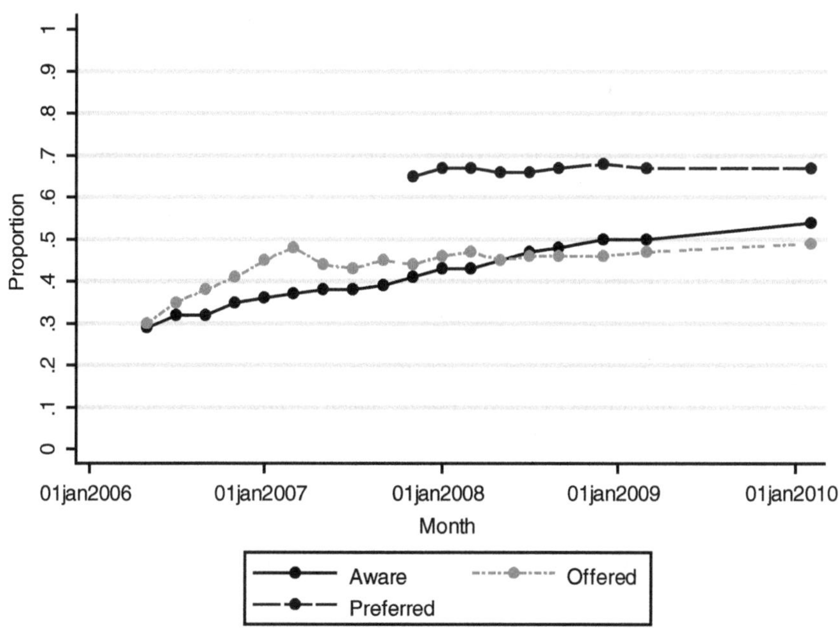

Fig. 1 Trend in patient awareness and availability of choice of secondary care provider in England 2006–2010 (Department of Health 2010)

their treatment location (40 %), which may indicate that either the patient did not seek to discuss options with their GP or that their GP felt there was no need given the nature of the patient's referral, the patient's willingness to accept their judgment or the uniformity of accessible services. Alternatively, the process of choice may help patients develop a preference for a certain provider. Patient choice appeared to improve the appointment process with 57 % of patients offered choice able to have a choice of appointment day and time compared to half that fraction for patients not offered choice (27 %). Moreover, 45 % of patients offered choice said they were very satisfied with their waiting time compared to 38 % not offered choice (4 % of patients not offered choice were not at all satisfied with their waiting time).

The offer of choice to patients increased their likelihood of being referred to an ISTC or private provider. The proportion of patients that were referred to an independent provider having been offered choice in 2010 was 60 % despite only 48 % being aware of choice prior to referral. The apparent increased likelihood for patients choosing an ISTC was related to shorter waiting times at independent providers as 22 % of patients choosing independent providers indicated the waiting time as determining their choice, compared to 9 % who chose an NHS provider. A significantly higher proportion of patients (17 %) were very satisfied with waiting times for their appointments at independent providers, and these patients were willing to travel further to be treated at independent providers. Significantly fewer patients treated at ISTC indicated closeness to home as the main factor determining choice (22 % compared to 38 %). Other factors were also important and it may be that patients perceived quality to be higher at independent providers. Pérotina et al. (2013) found differences in quality across certain dimensions between NHS and private ISTC. However, on average, no significant difference in overall patient reported quality was found after controlling for observable and unobservable differences in patient and provider specific characteristics. Dixon et al. (2010) found that few patients recalled being offered a private sector option (8 %) and only 19 % were aware before visiting the GP that NHS funded treatment in a private provider was an option, which may be explained by the limited set of treatments privately contracted for by the NHS.

The patient's local GP or other staff members at their GP practice were the single most important source of information for patients when deciding to choose their provider with 48 % of patients offered choice indicating this as their primary source. Patients also made extensive use of the experience of friends and family (33 %). There was limited use of the NHS Choices website or other internet sites (5 %), which may reflect lower internet usage among the elderly. The most frequently cited factors for choosing a particular provider were cleanliness and low infection rates (76 %), quality of care (65 %), length of waiting time (64 %) and friendliness of staff (57 %). The most influential single factor reported by patients offered choice was a hospital close to home or work with easier access and lower travel costs (43 %); followed by personal experience (18 %); perceived quality, reputation and cleanliness (11 %); and waiting times (10 %). Dixon et al. (2010) also found that one-fifth of patients would only consider their local provider as a treatment option in most circumstances. Patient choice may also have reduced the strength of the agency relationship between patients and their GP. In booking appointments, 46 % of patients offered choice had booked the appointment in the practice (either using the electronic Choose and Book system with the GP or the GP wrote a referral) compared to 67 % not using choice with 38 % of patients offered choice having used the Choose and Book call center and 12 % the internet.

2.2 Patient Mobility within the NHS

This section analyses the magnitude and trends in the extent of patient mobility as measured by all trips for health care (medical reasons), travel distances and travel journey times for a representative sample of the UK population. The analysis provides an insight into how patient mobility has changed and whether the policy of patient choice may have influenced patient mobility. In particular, if patient choice has resulted in patients traveling further or to different provider locations.

2.2.1 Frequency of Mobility for Health Care Utilization

The propensity of individuals to travel for health care reasons is measured by the frequency of journeys made by individuals for medical reasons. Data from the National Travel Survey (NTS) (Department for Transport 2004), conducted annually between 2002 and 2010 for a nationally representative sample of UK households was used to identify all trips made to a health care provider from the journey diary of each household member that recorded every trip made during a 7 day travel diary week with a destination located in England, Wales, or Scotland. Trips for medical reasons included GP consultations, hospital visits (A&E attendances, outpatient appointments or inpatient admission), and pharmacist visits to pick up prescription medication or consult for medical reasons, dental, optometry, physiotherapy, chiropody or any other medically relevant journey. The precise type of medical visit was not available in the publicly released data. However, from other sources (IPSOS/MORI 2012) the majority of health related visits in a year would be to a GP practice (approximately 83 % of the population made at least one visit per year), followed by specialist outpatient appointments (39 %), dental visits (39 %), trips to pharmacists (33 %), attendances at A&E (23 %), inpatient hospital admissions (18 %) and visits to an NHS walk in center (17 %). It should be noted that it is not always necessary to travel to receive health care as GPs can make home visits, there are home care services, a growing use of home telemedicine services (12 % of the population recorded using the free national telephone medicine service NHS Direct during the year) as well as long term care residents receiving health care. It is also possible to purchase pharmaceuticals and health care products over the internet or have pharmacies deliver prescriptions.

Results indicate that over time there has been a statistically significant increase in the probability of making at least one medical visit during a 1 week period from 0.13 (95 % CI: 0.124–0.136) in 2002 to 0.145 (95 % CI: 0.138–0.151) in 2010. On average, for individuals undertaking at least one trip in 2010, they made 1.30 trips (95 % CI: 1.27–1.33) for health related visits during that week. This is not just a reflection of increasing mobility in general because the percentage of total journeys for health care (which conditions on changes in general mobility) also increased significantly from 0.96 % (95 % CI: 0.91–1) in 2002 to 1.2 % by 2010 (95 % CI: 1.13–1.25). An indication of the magnitude of this change can be understood by estimating the average number of health related trips per person per year, which has increased by on average 1 additional trip per year from 8.7 trips in 2002 (95 % CI: 8.22–9.13) to 9.8 (95 % CI: 9.29–10.33) in 2010, an increase of 13.1 %. This implies a total number of health care journeys in 2010 for the population of 60 million in England, Wales and Scotland of nearly 590 million trips, which is an increase of over 60 million trips in a decade.

2.2.2 Travel Distances and Journey Times for Health Care Visits

The distributions and time trends in recorded journey distances and travel times to and from health care destinations were analyzed using the NTS surveys between 2002 and 2010. The utilization of most NHS services is free at the point of use; however, accessing care has indirect cost in terms of travel costs and the opportunity cost of time. The average reported single direction travel distance from the origin location of the trip to the health care destination for all health care journeys occurring anytime between 2002 and 2010 was 8.1 km (95 % CI: 7.9–8.3 km). The mean does not convey the highly right skewed nature of the distribution of travel distances, not unlike that seen with health care expenditure data, with a small number of individuals willing to travel very long distances for health care (most likely specialist care). The median distance traveled was 4.8 km, with 95 % of trips below 26 km. Only 1.4 % of trips for health care exceeded 50 km with 0.43 % over 100 km, and the maximum recorded internal UK journey was 433 km. The vast majority of trips are concentrated between 1 and 25 km. The relative frequency of long distance trips defined as being over 80 km or more in the population is quite low with between 0.08 and 0.11 trips per person per year which translates into around 4.7–6.6 million trips nationally.

Total travel time from the originating location of the journey to health care destination, and journey to the next location (which may be different from the location of origin e.g. a trip may have started at home, involved visiting the GP, and then traveling to the place of work) was on average 43.6 min. Again, the distribution was highly skewed to the right with the median round trip journey time of 30 min, with 90 % of all trips entailing between 10 and 115 min of travel with the longest 1 % of trips lasting over 210 min. Figure 2 shows the kernel density plot of the distribution of total round trip journey travel and visit times. An analysis of the total time commitment of all health care related visits, including

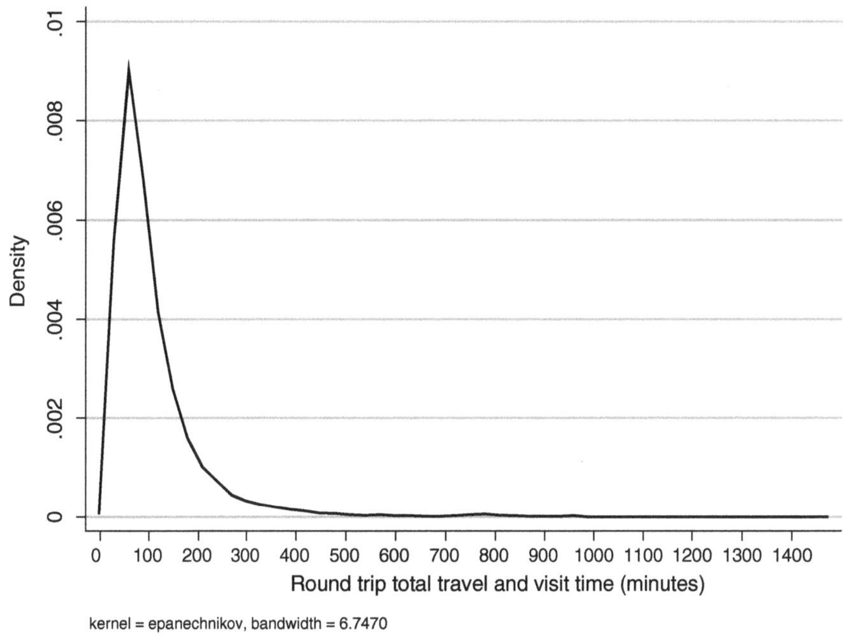

Fig. 2 Kernel density plot of the distribution of total round trip travel times for all individual trips due to medical reasons 2002–2010 (NTS 2010)

time spent attending the health care appointment (excluding inpatient overnight stays) found that on average, total health care day trips lasted 109 min, with median trip duration of 80 min. Additionally, 90 % of visits took between 25 to 270 min with 1 % of visits lasting more than 540 min.

Figure 3 shows that between 2002 and 2010 there has been a slight upward trend in average distances traveled for all health care related visits, although the annual differences are generally not statistically different. In 2002 the average recorded travel distance in the UK was 7.23 km (95 % CI: 6.71–7.74 km), average distances traveled were greatest in 2007 at over 1.5 km further [8.88 km (95 % CI: 8.09–9.66 km)] but had reduced to 8.11 km (95 % CI: 7.57–8.65 km) in 2010. Total time spent traveling and consulting for non-inpatient care has increased significantly on average over time with mean visit times of 101.7 min (95 % CI: 97.3–106.1) in 2002, increasing significantly to 115.99 (95 % CI: 110.56–121.4) in 2008, although this had reduced to 107.66 min (95 % CI: 103.5467–111.77) in 2010.

The increase in travel distances may be explained by the patient choice initiative in England and the waiting times database in Scotland to provide more information about waiting times to encourage usage of alternative hospitals. In Wales there was no policy to offer patients more choice of secondary care provider. At the same time there were also initiatives to improve access to primary care and specialty care in the community, which may have reduced travel distances

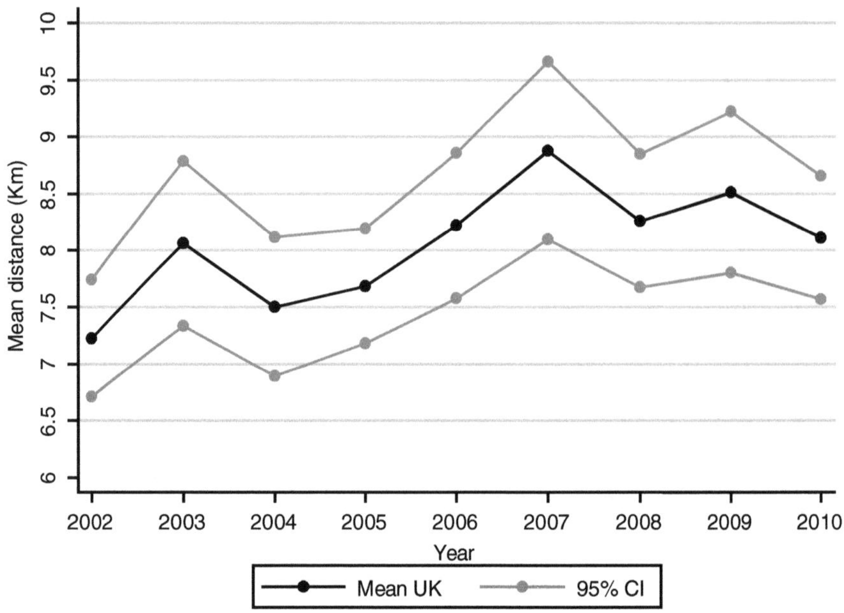

Fig. 3 Trend in mean distance traveled (km) for all health care related trips in the UK (excluding NI) 2002–2010 (NTS 2010)

(Department of Health 2008a, b). The NTS data covers all health related trips and therefore may not be specific enough to identify the effect of patient choice over time.

2.2.3 Cross Border Flows of Patients Within the UK

The devolution of policy making with greater autonomy for Scotland, Wales and Northern Ireland has lead to different health care policies across the United Kingdom. It is therefore of interest to look at cross border flows between England and Wales, England and Scotland and Scotland and Wales. An analysis of the NTS data between 2002 and 2010 found that no patients recorded travel for health care reasons between Wales and Scotland anytime in the weekly travel diary. Very few patients traveled between Scotland and England with only 0.12 % of total health care related trips (approximately 107 surveyed patients per year) with a destination in Scotland originating in England. Similarly, 0.12 % of health care journeys with a destination in England originated in Scotland (approximately 55 surveyed patients per year), hence a small net gain of patients for Scotland. This suggests either very little difference exists in the accessibility, quality and range of health care services between the two countries, so that Scottish and English patients have little incentive to cross the border for medical care. There is also a significant distance between the major specialist care and population centers in England and Scotland (for example the distance between Edinburgh and Newcastle is over 150 km) and historical sentiments between the English and Scottish could be a factor.

There were significantly more cross border flows between Wales and England with a greater net outflow of patients from Wales into England. Moreover, 1.29 % of all journeys with a destination in Wales originated from England (approximately 695 surveyed patients per year) and 2.19 % (approximately 1,293 of the surveyed patients per year) of all health care journeys originating from Wales had English providers as their final destination, hence there was a net outflow of 598 surveyed patient journeys each year from Wales or roughly 1 % of all health care journeys originating from Wales. There were significantly more cross border flows between Wales and England primarily because of a greater lack of services for Welsh or English patients around the border regions with around 17,000 Welsh residents registered with English GPs and 20,000 English residents registered with Welsh GPs. There are also fewer specialist hospitals in the Welsh border regions. The flows of more patients from Wales to England are likely a reflection of greater access and quality problems in the Welsh health service as greater investment and managerial pressure in England significantly reduced waiting times relative to Wales (Bevan and Hood 2006).

2.2.4 Cross Regional Flows of Patients in England Before and After the Introduction of Patient Choice

As well as looking at patient flows across country borders in the UK, inter regional flows between the nine Government Office Regions (GOR) in England were estimated combining data for the years 2002–2010. Analysis of inter regional flows in England showed that Greater London and the North East of England experienced the greatest net inward flows of patients with 1.15 and 3.3 % respectively gained above all patient journeys originating in those regions. Whereas Eastern England and South East England experienced a net loss of 1.5 and 1.14 % respectively, relative to all health care journeys originating in those regions, with the majority of these patients traveling to Greater London. The North East of England tends to gain patients from Yorkshire and the North West of England. The proportions of patients traveling outside of their region for all forms of health care is very small indicating that these patients are likely traveling for very specific treatments.

In order to investigate whether inter regional flows have been affected by the introduction of the patient choice policy, Table 1 shows a matrix of the relative distribution of patient flows across regions in the UK by originating and destination regions broken down by pre (2002–2005) and post (2006–2010) patient choice years. There appears to have been a small increase in the volume of patients traveling outside of their regions after the introduction of patient choice. The North East, Yorkshire and the Greater London regions experienced net inflows of

Table 1 Share (%) of the total number of patients traveling for medical reasons by region of journey origin and destination

Region (GOR)	Years	Patient destination shares (%)											
		North East	North West	Yorkshire	East Midlands	West Midlands	Eastern England	Greater London	South East	South West	Wales	Scotland	Total share destination
Patient origin shares (%)													
North East	02/05	4.61	0	0.022	0	0	0	0	0	0	0	0	4.63
	06/10	4.63	0	0.004	0	0	0	0	0	0	0	0	4.13
North West	02/05	0.006	12.58	0.043	0	0	0	0.007	0	0	0.008	0.013	12.66
	06/10	0.02	11.86	0.05	0.018	0.021	0	0.021	0	0	0.005	0	11.99
Yorkshire	02/05	0.053	0.017	8.09	0.047	0	0	0	0	0	0	0.011	8.21
	06/10	0.044	0.009	8.8	0.057	0.005	0	0	0	0	0	0	8.92
East Midlands	02/05	0	0	0.032	6.96	0.068	0.043	0.007	0	0	0	0	7.12
	06/10	0	0.02	0.095	7.18	0.11	0.076	0	0.005	0	0	0	7.49
West Midlands	02/05	0	0.074	0	0.007	9.07	0	0.047	0.007	0.008	0.044	0	9.26
	06/10	0	0.04	0.005	0.097	9.41	0	0	0.019	0.06	0.023	0	9.65
Eastern	02/05	0	0	0	0.04	0	9.95	0.26	0.046	0	0	0	10.29
	06/10	0	0	0	0.086	0	9.12	0.26	0.068	0	0	0	9.54
Greater London	02/05	0	0	0	0	0	0.14	11.27	0.2	0	0	0	11.61
	06/10	0	0	0	0	0	0.14	10.08	0.2	0	0	0	10.42
South East	02/05	0	0	0	0.017	0	0.041	0.28	13.55	0.087	0	0	13.97
	06/10	0	0.009	0	0.017	0.005	0.026	0.45	13.49	0.1	0	0	14.10
South West	02/05	0	0	0	0	0.051	0.007	0	0.067	8.73	0.05	0	8.90
	06/10	0	0	0	0	0.033	0	0.015	0.13	9.21	0.01	0	9.40

(continued)

Table 1 (continued)

Region (GOR)	Years	Patient destination shares (%)											
		North East	North West	Yorkshire	East Midlands	West Midlands	Eastern England	Greater London	South East	South West	Wales	Scotland	Total share destination
Wales	02/05	0	0.025	0	0	0.094	0	0	0	0	5.18	0	5.30
	06/10	0	0.068	0	0	0.027	0	0.022	0	0.005	5.49	0	5.61
Scotland	02/05	0.006	0	0.007	0	0	0	0	0	0	0	8.04	8.05
	06/10	0	0	0	0	0	0	0	0	0	0	8.76	8.76
Total share origin	02/05	4.6	12.7	8.19	7.08	9.28	10.17	11.87	13.87	8.82	5.28	8.06	
	06/10	4.19	12.00	8.96	7.46	9.61	9.37	10.85	13.91	9.37	5.53	8.76	

Comparison between pre patient choice policy years 2002–2005 and post patient choice policy years 2006–2010 (NTS 2010)

Table 2 Relative risk (%) of out of region travel for medical reasons compared to risk of inward regional travel between the pre (2002–2005) and post (2006–2010) patient choice policy years (NTS 2010)

Region of origin	Relative risk pre-choice	Relative risk post-choice	Change in relative risk
North East	−66.10	−93.91	−27.80
NW and Mersyside	−33.88	−8.05	25.83
Yorkshire	23.67	−25.05	−48.72
East Midlands	34.86	11.42	−23.44
West Midlands	−12.16	21.49	33.65
Eastern England	49.78	71.07	21.29
Greater London	−43.43	−55.73	−12.30
South East	32.73	43.73	11.00
South West	84.99	13.87	−71.12
Wales	17.01	223.02	206.01
Scotland	−48.75	0.00	48.75

patients of between 1.5, 0.45 and 4.13 % respectively above the share of total patient journeys originating in each GOR. This compares to 0.86, −0.24 % (net outflow) and 2.4 % prior to the introduction of patient choice. Also, 7.1 % of patients treated in London were cross-border inflows post choice compared to 5 % prior to choice. Greater London increased their flow of patients from the South East region, the North West, Wales and the South West. The North East of England, Yorkshire and the South West experienced increases in their net flow of patients of between 0.6 and 0.7 % relative to the proportion of patient journeys originating in these regions. Wales, Eastern England, the South East, East of England and the West Midlands all experienced net outflows of patients in the post choice years with their relative share of patients falling by at least 0.6 %. The estimated relative risks of a patient traveling outside of their GOR of origin for health care for the pre and post choice periods are reported in Table 2. They show comparatively large reductions in the relative risk of patients traveling outside the North East, Yorkshire and the East Midlands for treatments after the introduction of choice with the relative likelihood of patients resident in these GOR traveling across the regional border decreasing by between 23, 28 and 49 %, respectively. This indicates that the introduction of choice lead to half as many patients leaving the East Midlands as visiting the East Midlands for health care from other GOR. The West Midlands, by contrast, lost a significant proportion of patients to the East Midlands, with patient choice leading to over a third more patients leaving the West Midlands for health care than arriving.

In Wales, where patient choice was not introduced, there was a very large change in patient flows due to choice in England with a 200 % increase in the relative probability of patients leaving Wales for treatment. The large net outflow of patients was due to a reduction of patients traveling into Wales from neighboring English regions such as the North West, South West and West Midlands after the introduction of patient choice in England, and not an increase in patients traveling outside of Wales for health care. This may be explained by the fact that

English patients near the border may have switched to using English GPs instead of Welsh GPs who could not offer choice of provider or that choice in England lead to English patients to switch from using welsh hospitals to a preferred English one given the much greater choice set in England than Wales (Watkins 2012).

It is possible that the pattern of these flows reflects relative differences in the quality of services provided in certain parts of the country. For instance, very poor quality of care at the Mid-Staffordshire NHS Trust resulted in a national public inquiry into the causes of the failure in standards of care. This trust was located in the West Midlands GOR that experienced a relative increase in the proportion of patients traveling outside the region for treatment. The inquiry found a number of problems with the regulation of quality and standards of care in this region between 2001 and 2008 (Francis 2013).

2.2.5 The Impact of Choice and Market Competition on Patients' Location of Treatment

Kelly and Tetlow (2012) investigated the extent to which patients and their referring GPs have changed their treatment locations for secondary care in England after the introduction of patient choice in 2006 and the growing market penetration of ISTC. ISTC were originally introduced into areas having long waiting times or lacking capacity, but are now creating a much more competitive market for most NHS providers, with the number of different ISTC provider sites having increased from 20 to 161 providers between 2003 and 2010. They found that the proportion of patients attending their nearest NHS Trust decreased over time while those attending ISTC increased significantly. ISTC market share grew by 8.5 and 7.7 % in orthopedics and gastroenterology respectively, so that by 2010 there were 475,000 patient attendances. There was also a small increase in the use of non-local NHS Trusts. Patients were willing to travel more than 15 km from their nearest NHS Trust to be treated at ISTC providers with the average distance traveled for orthopedics increasing by 0.8 km between 2006 and 2010. The referral market became more competitive with patients and referring GPs using more providers. The market for elective care also became significantly less concentrated with the mean Herfindahl index for a GP practice falling from 0.71 to 0.61 (equivalent to distributing all referred patients equally between 1.63 providers). The mean number of different hospitals used by GP practices also increased from 12 to 18. Taken together, their results suggested that there has been a significant change in how GP practices and their patients choose their preferred hospital for treatment. However, there was also an increase in the variation in referral behavior across GP practices within the same PCT, suggesting a differential response by GPs to the choice policy.

3 The Influence of External Markets on UK Patients and the NHS

The globalization of the world economy implies that nations are highly interconnected and interdependent. The World Trade Organization (WTO) and its General Agreement on Trade in Services (GATS) along with increased use of regional and bi-lateral trade agreements has opened borders (Smith et al. 2009) with a reduction of barriers to trade. The rapid flow of financial and human capital, expertise and information, imply greater competition and access to global markets which has seen the growth of multinational corporations. Trade in health services has been affected by the same forces. Chilingerian et al. (2012) assessed that the world of health care has started 'flattening' with a levelling in the imbalances in health care provision. This has resulted in an increase in the flow of patients, medical practitioners, technologies, service providers, policy tools and investment capital between countries.

These forces are not a new phenomenon for the UK. The British Empire was a global system before "globalisation", and the Commonwealth of Nations and British missions contributed the "Anglobalisation" of the world economy with the exportation of British culture, language, institutions and values around the world (Peterssen and Osterhammel 2005; Ferguson 2003). The empire effect has been associated with a doubling in bilateral trade (Mitchener and Weidenmier 2008). The observed effect was associated with improved communication and transport; common culture and language; the rule of law and property rights; the establishment of currency and customs unions; and preferential trade agreements combined with guaranteed investment securities.

Britain as an island with numerous large serviceable airports, ports and connected road and rail routes, including the Channel Tunnel has been ideally placed to exploit access to European, American, African and Asian markets and attracts substantial foreign investment. The UK has a very multicultural society. Its legacy as an open, trading economy with a colonial past, lead to many waves of migration to and from former colonial countries, and more recently from members of the EU common market and EEA in particular new member states during the first decade of the twenty-first century.

Globalization is likely to have significant implications for population health and to create challenges for the financing and provision of national health care systems with direct and indirect effects through national wealth, economic growth, and access to more resources (Smith 2010). The relative efficiency of a nation's health care system is becoming an important source of competitive advantage, adding to productivity and well-being. However, the costs of health care are also placing a greater financial burden on households. National health systems are now subject to more international competition and markets for care provision are no longer the preserve of 'local' providers with greater international investment in private health care and a decreasing role for the nation-state (Kangas 2010). Globalization and the free movement of medical personnel combined with increased demand for health care and population pressures from aging and fertility decline in developed nations are creating imbalances in the distribution of the global health care workforce. Globalization also exacerbates externality health risks such as the rapid spread of infectious diseases (e.g. pandemic influenza, coranaviruses, drug resistant bacteria), exportation of carcinogenic pollutants in the air, water supply and over land as well as climate change (Huynen et al. 2005).

Following Lunt et al. (2013), the range of internationally mobile patients include: temporary visitors abroad who holiday or travel, but have to use health services as a result of an accident or a sudden illness; long-term residents abroad such as ex-pats choosing to retire in another country; patients from common border countries with cross-border health care agreements to provide cross-national publicly funded health care services from providers in each other's countries such as England and Wales or countries in the EEA; outsourced patients that opt to be sent abroad by health agencies using cross-national purchasing agreements; and medical tourists who are mobile through their own volition, paying either out of pocket or having private health insurance that reimburses planned treatment abroad. The distinction between medical tourists or cross-border patients in the EU context is a gray area because EU citizens can use their rights to access medical care directly, having to search for, organize and finance their treatment abroad in a similar way to medical tourists, later claiming coverage from their domestic third party payer (Bertinato et al. 2005). The perspective is more general than others such as Glinos et al. (2010) who focus on patients crossing national borders traveling primarily for the purpose of obtaining health care services in their destination country.

Exporting health care nations are those countries attracting inbound patients. Their pull factors include comprehensive and advanced care; highly focused and specialized care; better quality; less costly; and faster access to innovative technology as well as better information through marketing and international accreditation providers. Importing nations send overseas patients with the main push factors being the lack of availability of treatment at home; ability to search; knowledge of other countries; experienced user networks or professional brokers; ability to pay and travel as well as willingness to leave local support networks (Lunt 2011). Loh (2013) analyzed health related travel expenditure data from national Balance of Payments accounts from 2003 to 2009 and found health care expenditures attributed to foreign travel increased significantly, but only between nations with higher existing volumes of health travelers. The rise was due to an increase in volumes of travel and not expenditure per traveler.

3.1 Outward Patient Journeys from the UK to Exporting Countries

There has been a recent change in the characteristics and direction of flow of patients traveling for health care, with a relative increase in patients traveling from developed countries to emerging countries who have specialized in providing quality health care at comparatively lower cost (Deloitte 2009). These patients are from countries where health care is expensive with increasingly incomplete financial coverage, such as the United States, Switzerland, and Japan as well as health systems with fixed budgets, capacity shortages, long waiting lists and rationed services, which delay adoption of innovative new technologies such as Canada, UK and Germany. These countries are likely to see more patients treated abroad in future.

For patients in the UK there are significant push factors. NHS care is rationed due to the closed funding of care, which can reduce the set of available treatments on the NHS, delayed adoption of innovative treatments and technology, and capacity shortages that lead to significant waiting times to access specialized treatment. The UK private health care sector is also comparatively more costly than abroad in particular dentistry, cosmetic surgery and fertility treatments that are not always routinely covered on the NHS. There are also, however, a number of factors that increase the inertia of patients to seek treatment outside of the UK. NHS care is free at the point of use, and provision relatively locally distributed with financial resources allocated to a significant extent on the basis of population health needs. There is also access to a domestic private sector which contracts with the NHS to provide new services or can be directly accessed by privately insured patients or those willing to pay.

3.1.1 Public Satisfaction and Comparative Performance of Health Care in the UK

To gain an understanding of the push factors that may drive UK residents to seek treatment abroad it is possible to look at some trends in subjective as well as objective measures of the performance of UK health system relative to other European and OECD countries. Public satisfaction in the NHS in England, Scotland and Wales has improved significantly in the last decade. From an all time low of between 30 and 40 % in 1997, it increased to above 65 % by 2009 (see Fig. 4). This is likely to reflect the increased investment and managerial incentives that have lead to significant reductions in patient waiting times (Appleby 2012). Following austerity policies that have capped the level of expenditure on the NHS, and demanded efficiency savings of up to 20 % of expenditure, along with a small rise in waiting times, satisfaction fell significantly to less than 60 % in 2011. Satisfaction among service users remains quite high. A 2012 survey of 1,015 members of the general public (IPSOS/MORI 2012) indicated that 82 % of

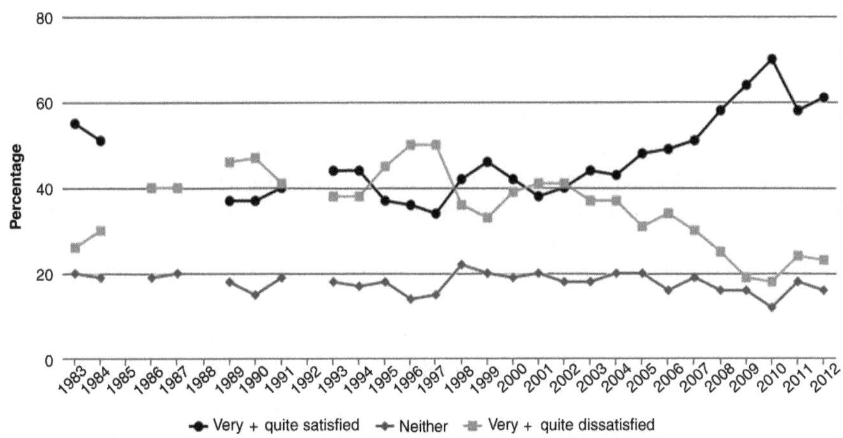

Fig. 4 Patient Satisfaction with the UK NHS 1983–2011 (Appleby 2012)

individuals who had visited a hospital in the previous year thought their local NHS was providing them with a good service, while 13 % disagreed and 5 % were indifferent or uncertain. Loyalty to the NHS is quite strong with 71 % of individuals having considered the NHS to be one of the best health services in the world. Views about safety and quality of care, however, were less confident with 24 % of individuals not having felt safe in an NHS hospital and only 37 % of respondents thought the NHS puts quality at the heart of everything it does.

Waiting times and unmet need have been identified as one of the most important factors influencing UK patients' willingness to travel for health care. Around 2.14 % of individuals in the UK reportedly had an unmet need for medical care because of waiting lists in 2005; countries with similarly high proportions include Lithuania (2.32 %), Poland (2.26 %), Estonia (2.25 %), Germany (1.74 %) and Latvia (1.72 %). There was significant unmet need for dental care as a result of waiting lists with 4.1 % of the population having reported problems in accessing a dentist. This was not a problem in other European countries (Busse 2011). Official data has shown significant reductions in patient waiting times for specialist treatment. The median referral to inpatient admission time was 14 weeks in 2007 which had fallen to 8.4 weeks in 2012. There is also much less variation in waiting times with 95 % of patients treated within 14.8 weeks compared to nearly 50 weeks in 2007 (Nuffield Trust 2013). Despite significant reductions in reported waiting times, only 31 % of individuals having used hospital services in the last year believed hospital waiting times for planned treatments were getting shorter with 49 % disagreeing and 17 % unsure. Moreover, 20 % of the surveyed population in the NHS survey considered long waiting lists as a problem (IPSOS/MORI 2012).

The UK's relative health system performance across a range of indicators appears to lag behind other European and OECD countries. UK rates of mortality from conditions amenable to health care were 83 per 100,000 in 2007, 50 % higher than in France, and a third greater than in the Netherlands and Sweden. The UK ranked 18th and 20th out of 31 OECD countries for female and male age adjusted amenable mortality rates (Gay 2011). Health adjusted life-expectancy of 72 years in 2008 also ranks in the bottom third of OECD countries, although measures of mortality may be capturing underlying differences in the severity of population health status and other factors not under the influence of the health care system. Age and sex adjusted acute myocardial infarction (AMI) mortality rates are often used as a proxy measure for overall quality of specialist and preventive care services and the UK performs better with rates 19 % lower than the OECD average in 2007 (OECD 2009). Quality of cancer care, however, appears to be relatively poor. Between 2002 and 2007, 5 years cancer survival rates in the UK were 10 % below the OECD average despite comprehensive screening program. Breast cancer 5 years survival was 15 % below that of the United States and 5.5 % below France. Male 5 years survival for colorectal cancer in the UK was 16 % worser than the Netherlands.

Across nine inpatient and maternity patient safety indicators collected and analyzed by the OECD, the United Kingdom ranked ninth out of 15 countries on average despite having only the sixth worst level of co-morbidity. After adjusting for co-morbidity, the UK was ranked 12th out of 15 OECD countries just in front of Canada and the United States, but behind the leading countries of Denmark, Belgium, Singapore, Spain, Italy and Finland (Drösler 2009). Surgical errors and unintended lacerations and punctures were close to the OECD average, but error rates were for example 35 % lower in Denmark (OECD 2010). Length of stay in hospital was on average 7.2 days, a day above the OECD average. Postoperative pulmonary embolism and deep vein thrombosis rates were significantly higher than the OECD average although sepsis and catheter related infection rates were just below average. The UK also ranked 12th out of 34 European countries according to its scores for performance across 42 indicators covering six dimensions of health care system consumer orientated care, including patient's rights, accessibility, prevention, quality of care, and prescribing (Björnberg 2012).

In the UK since the Bristol Royal Infirmary Inquiry in 2002 into failings of pediatric heart surgery services (Kennedy 2001), regulators have attempted to improve the monitoring of patients' outcomes at hospital and later GP practice level. The Doctor Foster organization, National Institute for Cardiovascular Outcomes Research, the national quality of care regulator the Care Quality Commission (CQC) and numerous other organizations have been collecting analysing and disclosing patient outcomes data across NHS providers for the last decade (Bridgewater 2013; Foster Intelligence 2012), including standardized mortality rates for hospitals and certain specialties, hospital readmission and infection rates, inspection outcomes, patient satisfaction and recently patient reported quality of life outcomes before and after routine elective surgery (Smith and Street 2013).

The disclosure of data as well as NHS regulatory and clinical governance regimes proved ineffective in preventing the persistence of terrible standards of care at Mid-Staffordshire NHS Trust, which was a complete systems failure from health care professionals, management, regulators and policy makers with patients' concerns not acted upon (Francis 2013). Further concerns about quality of care at 14 more hospitals with excess standardized mortality rates have lead to their thorough inspection and review of performance, which identified serious problems with patient care related to staff shortages, motivation, and quality of management. It also highlighted the need to improve co-ordination and analysis of the variety of data collected, but not routinely shared or interpreted (Keough 2013). The most serious problems were in managing emergency admissions, particularly on weekends and late at night, which depending on the nature of the emergency may not be circumvented by greater patient mobility.

3.1.2 Willingness to Travel and the Factors Influencing the Propensity to Seek Treatment Further from Home

Given the above evidence there would appear to be a number of reasons to justify a UK patient's decision to travel abroad for treatment. In 2002 the British Medical Association (BMA) surveyed 1,982 adults aged over 15 years and asked how far people would be prepared to travel for treatment if they faced a long wait on the NHS and if the NHS would pay for their care (Beecham 2002). Overall, 42 % of patients were willing to travel outside the UK for treatment (15 % to Europe and 27 % anywhere in the world). For the 58 % not prepared to travel abroad, 7 % would only go to their local hospital, while 47 % would travel 50 km or more to obtain shorter waiting times. Attitudes to the provision of NHS health care by the private sector were mainly positive with 76 % believing care provision would be the same or better.

In 2007 IPSOS/MORI undertook a nationally representative survey of 2,028 individuals aged over 15 in 2007 and a small number of qualitative focus group interviews about patients' attitudes to local provision of cancer services, in particular attitudes to travel to obtain better quality of care (IPSOS/MORI 2007). Figure 5 plots respondents' reported willingness to travel for improvements in different dimensions of care. A similar percent (40 %) of people to the BMA study stated they were willing to travel any distance for faster and better quality treatment, in particular better outcomes such as 5-year survival rates. There was a clear bi-modal distribution with less than 45 % prepared to travel for more than 2 h for treatment across 6 different quality dimensions. A higher proportion of respondents indicated they will travel further for reduced 30-day mortality, but most would not be prepared to travel further solely for better information, to obtain shorter length of stays or better standards of care, believing these either to be of less value or something that should be uniformly provided no matter where they are treated. The majority (73 %) or respondents were willing to travel further than their local hospital for at least one type of service improvement. Two-fifths of respondents said they would be more willing to travel if they could be treated quicker. Fewer (27 %) responders were willing to travel further than their local hospital for cancer treatment involving daily trips with side effects for a 6 week period.

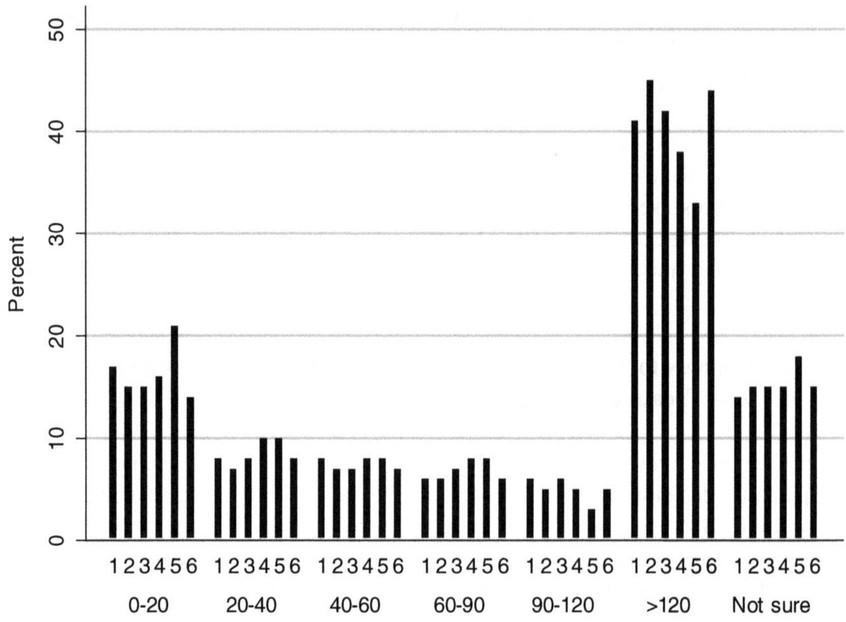

Fig. 5 Distribution of the willingness to travel in minutes for improved dimensions of cancer treatment for survey respondents in England, 2007 (IPSOS/MORI Cancerbackup study 2007) Category key: *1* Length of stay of 4 instead of 10 days; *2* lower 30 day mortality; *3* lower risk of complications; *4* Better standards of care *5* better information; *6* better 5-year survival. Travel time categories expressed in minutes

In order to place the travel time responses in context of the distribution of UK hospital travel times, the NTS indicated that in 2007, 36.3 % of households had access to a hospital within 20 min, 54.4 % had to travel between 20 and 60 min to their nearest hospital while 8.9 % had to travel more than 1 h to their local hospital. Hence, a significant proportion of people were not prepared to travel more than 1 h which represents the maximum journey time to the closest hospital for 91 % of the population.

The willingness to travel further was greater among individuals from higher social classes (professionals, managers, teachers, individuals with responsibility and service sector workers) who were more willing and able to travel for better treatments as they had the financial means, access to better information, and higher expectations about the possibility of obtaining improved treatment elsewhere. Certain subgroups were more likely to travel, particularly women over 55 years. Although the elderly in general indicated greater reluctance to travel 60 % of individuals aged 65 would be prepared to travel any distance, given a particular circumstance. For many traveling further for treatment was not an option with 8 % indicating they would never travel beyond their local hospital and the qualitative study revealed that individuals would not be satisfied with a situation of 'reasonable' cancer treatment provided at local hospitals. The benefits of improved local services were considered significant and many thought they outweighed the cost of further travel as people placed a very high value on being able to be treated in close proximity to family for social and emotional support in a familiar environment. Local treatment was favored by those having dependents and work commitments. Interestingly, part of the desire to remain local was mistrust that services could be better elsewhere due to a lack of information and scepticism about statistics regarding the quality of services or reported patient outcomes elsewhere. Hence, their expectations about better outcomes from an alternative hospital appeared to be lower than the costs of traveling and having to be further from friends and families.

Interviews revealed variation in the quality of health services. Residents in Kettering in the West Midlands felt they received a poor quality of health care service and traveled outside the local area to receive better care (which corroborates patient flow evidence from the NTS). Below 'reasonable' care standards were also a view of Kidbrooke residents in South East England. First hand experiences of cancer services were very mixed with a belief among some that the NHS had prioritised cancer services, attempted to speed up access and do everything possible to support patients, while others reported poor communication, service levels and treatment. Most patients expressed a desire to be fully informed about all options and wanted self-determination in treatment plans. There was support for the idea of creating centralized specialist services for cancer, which could involve increased travel. However, access to a car or public transport or financial barriers were factors that affected willingness to travel. Concerns were also expressed about equality of access if services were centralized as it would be harder for some people to access these services. Also in real situations numerous other factors need to be taken into account, particular severity of the cancer, nature of the treatment, continuity of care and personal circumstances.

3.1.3 Magnitude and Characteristics of Patient Mobility of UK Residents to Health Care Providers Outside the UK

Precise data on the number of patients leaving the UK for health care reasons is not systematically collected. Lunt et al. (2013) provided estimates from the 2010 International Passenger Survey [Office for National Statistics (ONS) 2011] of a total of 63,000 UK patients who traveled overseas for health care, spending around £62 million funding their treatment. This represents approximately 0.11 % of total annual overseas journeys. Figures of between 50,000 and 100,000 patients per year have also been reported (Youngman 2009), of which 75 % of trips were for cosmetic surgery. Other prevalent medical reasons include: dentistry, cardiology/cardiac surgery, orthopedic surgery, bariatric surgery, fertility treatment, organ transplantation, eye surgery, diagnostics, and more recently tissue and stem cell implantations (Lunt et al. 2013). Between 2000/1 and 2004/5 UK expenditure on reimbursements for planned EU cross-border treatments increased from 26.6 million to 49.5 million in 2004/5 (European Commission 2006). Also, 5,172

E112 agreements for planned cross-border treatment by UK resident patients in other EU countries were approved.

For patients unable or unwilling to access NHS elective treatments traveling can achieve significant cost savings compared to private treatment in the UK. Procedure costs (excluding travel and hotel charges) for cosmetic surgery abroad can be between $1,000 (£657, €756) and $3,000 (£1,972, €2,269) less in India and Cuba compared to UK private sector prices (March 2011); dentistry costs for implants are nearly $1,000 (£657, €756) cheaper in Hungary and hip replacements $4,500 (£2,958, €3,404) less. Knee replacements are also relatively cheap in India costing around $3,000 (£1,972, €2,269) less and hip replacements in a 66 % cost saving. Heart bypasses in India are £10,000 (£6,574, €7,563) less than UK private sector prices and cataract operations cost £60 ($91, €70) compared with around £3,000 ($4,564, €3,516) privately in the UK (Lunt et al. 2011, 2013). For example a case reported in 2001 in the UK press (Wilson 2001) of a 70 years old man from Tamworth in Staffordshire who paid £5,000 ($7,749, €5,861) in total for very satisfactory treatment lasting 6 days in South Africa instead of waiting up to 18 months in the UK at a saving of £3,000 ($4,564, €3,516) compared to UK private treatment. The South African clinic had treated around 40 UK patients a year.

The majority of UK citizens traveling abroad for health care treatment appear to be doing so for cosmetic surgery and dental treatments (Boseley 2013a). This has been facilitated by the growth in cheaper air travel and the internet (Lunt et al. 2010; Medekher 2013). Patients are able to save between 40 and 70 % of the price of UK treatment with flights and accommodation included. Purely cosmetic treatments are not covered by the NHS, but bariatric surgery can be performed on the NHS, although waiting lists are very long (up to 2 years) and criteria for funding vary considerably across local PCT. The most popular destinations for UK medical tourists appear to be the Czech Republic (for bariatric surgery or fertility treatment); Belgium (for breast augmentation and bariatric surgery); Poland, Hungary and Tunisia (for tummy tucks, liposuction, face and breast operations); and to Budapest or Malta for dental treatments. Treatment links between the UK and some countries exploit longstanding historical and cultural connections or more recent waves of migration from Eastern Europe. For example, there are trade agreements between the UK and Malta (Muscat et al. 2006).

Holliday et al. (2013) interviewed 103 patients, 43 of whom were British, traveling abroad for health care, and found that patients mainly relied upon recommendations from friends and social networks and internet resources rather than hard evidence to choose their destination of treatment. Some individuals do extensive research such as checking qualifications, accreditation and what local people think of providers. There are also UK based agencies and brokers who can organize treatments with approved providers (Boseley 2013a, b). Most individuals were women with modest to medium incomes employed in the service or professional sectors including administrators, nurses, hotel porters, hairdressers, students, police officers and teachers. British patients minimized their trip durations to between 5 and 7 days in order to save costs. Their main reasons for traveling abroad for their treatment appeared to be value for money, and they regarded their treatment as an important investment that added value to their body, appearance and status. It was also the only way many could afford to change something with their body they were unhappy about as the treatment was unavailable on the NHS or there were very long waiting times. Patients also believed that the NHS should not pay for their treatment and were put off by the high price at UK private providers as well as the perception they were more financially motivated than medically concerned. Most of the patients they spoke to were satisfied with their surgery abroad.

Culley et al. (2011) interviewed 51 people in the UK between May 2009 and June 2010 and analyzed patient motivations for traveling abroad for fertility treatment. Fertility treatment is recommended by NICE for NHS funding but there are restrictions based on age, and previous children or attempted cycles as well as significant waiting lists. The main reason given by patients for traveling abroad was a desire for timely and affordable treatment in over two-thirds of cases (71 %). The most common single factors were: donor shortages in the UK (37 % cases); cost (25 %) which can average around £5,000 ($7,606, €7,749) a cycle in the UK; perceived better success rates abroad (24 %) partly related to a desire for multiple embryo transfer as well as unsatisfactory care in the UK (14 %). There was also the attraction of a holiday setting with a less stressful environment that can prevent disclosure of treatment to employers or family. Patients traveled to 13 different countries with Spain and the Czech Republic the most popular. A high proportion (44 %) organized the trips with little assistance from UK medical professionals and some patients encountered unwillingness from UK professionals to assist them. The women had predominantly professional or managerial backgrounds and were paying out of pocket for overseas treatment. The mean age of women seeking treatment were 38.8 years and at the time of the interview 51 % of the women were pregnant with a multiple pregnancy rate of 19 %.

A recent case study estimated the potential gains from a bi-lateral trade agreement between the UK and India to treat NHS patients (Smith 2011). The colonial connection between the UK and India has already encouraged a medical market between the two countries with members of Britain's Asian communities traveling back to utilize Indian health care, which has started to expand to all sections of British society. The Indian Government was one of the first emerging markets to realize the potential economic benefits from investing and incentivising health care tourism and has actively promoted it as a desired export. Policy initiatives include the use of lower import duties, prime land at subsidized rates and tax concessions. Additionally, the Indian Government has introduced a special type of visa for medical tourists, the "M", or Medical Visa. The benefits are seen to be foreign revenue for the health sector and economy and a reduction in physician exportation to the US and UK whose costs of training are subsidized by the Indian Government (Wilson 2001). These policies, however, may divert much needed resources from the domestic public health system, to the benefit of private hospitals. India has around 20 % of the total market share among major Asian providers, namely Thailand, Singapore and Malaysia. Between 2009 and 2011, medical tourist arrivals in India grew by 30 % and estimates of 500,000 medical tourists are forecast for 2015 (Medekher 2013). Indian physicians speak English, are well trained with many Indian doctors qualified in the United States and providers are very productive with virtually no waiting time. Moreover, 17 hospitals have received the Joint Commission International (JCI) accreditation who indicated that health care in India is as good as or better than US care. The study estimated substantial cost savings to the UK NHS from sending its patients to India of between £120 and £200 million as well as indicating the savings from reduced waiting lists. Hence, there would appear to be significant potential gains of using India as a provider of health services to the population of the UK. One example is an agreement signed in 2003 between the British private health insurer BUPA and Ruby Hospital in Kolkata.

3.1.4 Safety, Regulation and Ethical Concerns

Lunt et al. (2011, 2013) review the risks of traveling abroad for treatment and evidence of adverse events experienced by UK patients. There are inherent risks in medical tourism due to the uncertainty of the efficacy of treatments and the problem of information asymmetry between the provider and patient. These factors are exacerbated when seeking treatment abroad given the treatment may not have been fully researched by the patient, and a referral may not have been obtained from a knowledgeable third party acting as the patient's agent such as a general practitioner or local health authority. International accreditation and regulation of organizations can help overcome some of these problems, but false advertisement, unproven procedures, variability in the skills of individual physicians or quality of medical technologies are pervasive in health care (Wennberg 2011). There may be additional problems from post-operative care following treatment as standards may be worse when abroad due to a lack of continuity of care, missing health records and information as well as unfamiliarity with procedures or medical devices used by foreign physicians.

Traveling has inherent health risks, which are likely to be increased in patients undergoing treatment. There are risks of infection from bacteria and viruses in which travelers are not accustomed, particular in tropical climates with a different disease ecosystem. These carry not only increased personal risk but significant public health externalities as well (Kumarasamy et al. 2010; Hall and James 2011). There are studies documenting the rates of complications and adverse events arising following treatment abroad, although these studies generally do not compare patients to a control group of similar patients treated in the UK to determine the relative increase in risk. With regard to cosmetic surgery, 37 % of the 203 members of the British Association of Plastic, Reconstructive and Aesthetic Surgeons who responded to an association survey had seen patients in the NHS with complications arising from overseas cosmetic surgery (66 % response rate) such as severe wound infections and tissue loss. The survey found that 14 % of members had seen nine or more patients in the past year alone with problems as a result of cosmetic surgery abroad and a third of the surgeons had carried out much more repair work over the last 5 years (Jeevan and Armstrong 2008; Bosely 2007). The main areas for concern were aftercare, language barriers that may mean some patients do not understand what they are told or cannot communicate their own needs, lower quality standards and knowledge of care, abilities and experience of the surgeon. Another audit of consultants found 60 % had dealt with complications from treatment abroad with the majority emergency cases (66 %) (Birch et al. 2007). Of the 103 patients researched by Holliday et al. (2013) 16.5 % said they experienced complications and 8.7 % needed further treatment for problems such as infections when they got home. Another observational study conducted from 2007 to 2009 on patients identified 19 patients presenting with complications from cosmetic surgery tourism, mostly from Europe or Asia for breast augmentation procedures at a plastic surgery practice. Eleven of these patients were reported to have received NHS treatment, at a cost of £120,841 (Miyagi et al. 2011).

Kidney patient survival and function were concluded to be good following transplantation in Pakistan of patients resident in Scotland. However, there were serious infection complications and acute rejection (Geddes et al. 2008). A case-control study found significantly increased risk of infection, graft loss and mortality for organ transplantation of Asian patients from the Indian sub-continent with renal failure transplanted commercially abroad (Krishnan et al. 2010) compared to local transplantations. In terms of the impact on NHS maternity services from fertility treatment abroad, 26 % of high order multiple pregnancies were the result of fertility treatment performed overseas (McKelvey et al. 2009).

There are also important ethical issues, particularly relating to the trade in sourcing donor organs for transplant surgery in developing countries as well as use of embryos in fertility treatments. There were revelations that a fertility clinic in Spain offered patients the option of using embryos "left over" from previous treatments without the donors' explicit consent. In 2004, 114 of the 317 British couples treated at the clinic did not give consent, and their embryos were used to conceive children without the donors' knowledge, which runs against the UK law (Ireni-Saban 2013).

3.1.5 Case Study: UK NHS Outsourcing to Foreign EU Providers

Lowson et al. (2002) evaluated a Department of Health funded and supported pilot project to assess the implications of sending patients abroad for treatment. Concern about long waiting times for patients in the South East of England in 2001 led to the investment in three pilot projects by local HA to contract for elective surgical care with overseas health care providers. This followed the July 2001 European Court of Justice ruling entitling European patients to receive treatment within the EU funded by their local health care payer.

The three commissioning organizations were East Kent HA (EKHA); Portsmouth, Isle of Wight & South East Hants HA (P&IoW); and West Sussex & East Surrey HA (WS&ES). The pilots intended to offer patients the choice of overseas treatment to reduce waiting times for treatment in their local health economy. Given easy access to France via the Channel Tunnel or nearby ports as well as proximity to Gatwick and Heathrow airports, the pilots commissioned care services in France (Lille) and Germany (Essen and Cologne). P&IoW commissioned with seven German providers through German Medicine Net and Guy's & St Thomas's Corporate Development Team (Hanover, Osnabruck, Bielefeld, Hamburg, Damp and Celle). There was an informal arrangement whereby WS&ES and P&IoW patients could also send patients to Lille for treatment. Between July 2001 and July 2002 a total of 190 patients had been treated overseas from the 3 HA, 57.4 % in France and the remainder fairly equally distributed across the German sites.

The number of patients treated in Germany was significantly lower than anticipated and represented only 3 % of all the patients on the waiting list for treatment and 5 % of those waiting more than 12 months in December 2001. The lower than expected patient numbers were attributed to delays in setting up hospital contracts; conservative selection of patients from a limited range of specialties (mainly major joint surgery and cataracts) as the pilot sites decided to deal with ENT, urology and general surgery cases by finding capacity in local private facilities or neighboring NHS providers. NHS clinician collaboration was problematic and the earmarked budget limited treatment to no more than 250 complex cases with longer lengths of stay, although this number was not reached.

Patients were selected for travel based on long waiting times for treatment, absence of co-morbidities (cardiovascular, diabetes, epilepsy, frailty, obesity, and mental health problems) as well as intangible criteria such as a positive attitude to travel and general well-being. Patients were either identified as potentially eligible directly from the waiting list or attracted using publicity in the local media. Applicant attrition rates were high. Of the 214 patients who put themselves forward in P&IoW, only 40 finally received treatment in German hospitals (81 % attrition). Attrition rates in EKHA were between 90 and 50 % depending on who discussed overseas arrangements with the patient. It was found that patients were more likely to travel if they had been contacted by enthusiastic members of the overseas treatment organizational team knowledgeable about arrangements. GPs, clinicians and managers opposed or indifferent to the initiative were able to dissuade patients from traveling.

A pre-assessment specialist clinic was held in the UK prior to travel approval, and the pilots found that bringing clinicians from the EU hospitals to run the clinics in the UK worked well, enabling EU clinicians to meet local clinicians and to know patients before arrival. Clinical standards were very high across all hospitals due to effective research and selection using high quality criteria (large surgical volumes, accreditation, and top ranking among hospitals nationally) and knowledge of local care providers by intermediary organizations.

The main problems faced were non-clinical: travel and logistical arrangements (tickets, timetable, documents and information, ensuring patients' notes, consent forms and X-rays were to hand) as well as management of patients' care while in hospital (food, access for relatives, cultural differences, language and communication, overall patient attention and well-being). The pilot evaluated patient experience through blinded patient questionnaires with an 89 % response rate. Patients treated in France were on average 65 years with 53 % female. German treated patients were 59 years on average with 58 % female. Around 90 % considered longer waiting times for treatment in England as the main reason for wanting overseas treatment. Some patients found it useful to discuss having treatment overseas with a third party such as friends or relatives (24 % of French treated patients and 36 % for German treated patients). The main reasons for not wishing to travel overseas were arranging for care of children, relatives and pets; and organizing time off work. Moreover, 73 % of patients destined for France reported no concerns before treatment, but a higher proportion of patients 56 % had concerns about visiting Germany, including a wish to have been notified earlier of the decision to treat overseas; the ability to understand doctors and nurses; and lack of details about the journey, operation and post operative arrangements.

Less than a third of patients sent to France and 17 % of patients sent to Germany were actually accompanied by a relative or companion, which may be due to the fact that the patients traveled in groups with a local organiser. Patient satisfaction with care was very high with over 90 % of patients finding hospital staff helpful and courteous, rooms pleasant and at least 87 % finding information about treatment helpful. Perhaps unsurprisingly, about 80 % of the patients sent to France rated their food quite or very pleasant, compared to 49 % of patients treated in Germany! Overall satisfaction with the entire experience was high with 80 % of patients reporting they were very satisfied, although 24 % in France noted problems with language and communication. Suggested improvements included access to English TV, newspaper, radio and an interpreter; communal meeting places for patients; better food (Germany); and improved travel arrangements (less walking and waiting). Patients' remarks indicated the quality of treatment equaled or exceeded that of their local NHS service, especially physiotherapy and co-ordination of rehabilitation, and patients wanted better rehabilitation and aftercare arrangements on return to the UK, indicating problems with continuity of care as well as access to physiotherapy in the NHS. Patients were asked if they would be willing to accept treatment from alternative providers: 30 % of patients that traveled abroad would be happy to be treated in Spain and 25 % in Greece, 75 % would accept treatment in a UK private hospital and 84 % in another NHS hospital.

The views of local NHS stakeholders were more mixed with concerns raised about the cost-effectiveness of the initiative, the opportunity cost of not investing in the local health economy, and the limited impact on waiting lists given the organizational costs. However, from a patient's perspective the initiative reduced waiting times for some long waiters who received successful treatment without complications. There was a belief that treatment in France was a legitimate option to offer patients given the ease of travel and the value that patients felt from having a genuine choice. There were concerns about the undesirability of offering treatment to patients necessitating long journeys, having different clinical teams undertaking surgical procedures and longer term patient management.

There was little support from local specialist clinicians and GPs and possibly a deliberate attempt to restrict the number of patients deemed eligible to be treated abroad. One NHS Trust was very opposed to the project, refusing to participate in pre-assessment clinics and the local GP medical committee voted not to support the scheme on the grounds that funds should be spent within the local NHS health economy and was not value for money to the taxpayer. Clinicians may have resisted the challenge to their methods of working and management of waiting lists as well as the additional effort to co-ordinate care with overseas providers. They raised issues of a clinical nature, however competition from foreign providers and reductions in waiting times would threaten their income from private practice as well as their influence on the local health economy, hence imposing strict criteria for patient selection, not assisting in identifying eligible patients and discouraging travel in order to limit the extent of the project may have been a strategic response based on self-interest.

The main lessons from the pilot were the need to simplify the contracting and organizational process, and improve the cost effectiveness of the commissioning of care. This could be achieved by reducing transaction costs from making contracts with a few large volume, high quality providers, performing specialty care procedures that may require a longer length of stay and rehabilitation for example, and joint replacements. Concerns were raised about contracting at short notice with overseas providers with available spare capacity as this may imply poor quality of care. Procedures that can be done as day cases in the UK were too expensive overseas and would not be cost-effective. Other recommendations were the extension of the case mix and inclusion criteria to increase volumes; improved local arrangements for patient selection; improved patient travel and more local buy-in to remove administrative and clinical barriers.

This scheme was subsequently extended and an additional 269 patients were treated at a cost of £770,000 (£2,862 per patient compared to £5,789 per patient in the pilot). In 2002, the UK Department of Health set up the NHS Overseas Commissioning Scheme to facilitate treatment abroad. Another two pilots for orthopedic treatment overseas were established between April 2002 and March 2005 at seven Belgian hospitals identified as offering good quality treatment at cost-effective rates. One was part of the London Patient Choice Pilot (LPC) initiative, which contracted with five Belgian hospitals up until March 2007. A total of 432 NHS patients with hip and knee problems and 21 cardiac patients were treated in the five contracted hospitals between March 2003 and November 2004. However, no more patients were sent after September 2004, as the LPC budget was used up and more capacity became available in England. In total, around 300 patients a year were treated abroad at a cost of £5.7 million (£6,400 per patient). The schemes ended in March 2005 possibly due to their apparent high cost and availability of capacity provided by ISTC (van Ginneken and Busse 2011; Glinos et al. 2006). Hanna et al. (2009) conducted a retrospective matched case control study of 22 orthopedic patients sent to Belgium in 2003 with a group of locally treated patients. Self-reported patient outcomes were significantly better in the locally treated group, with four cases of complications in the overseas group compared to one in the locally treated group. The overall treatment cost for the patients sent to Belgium was £180,000 compared to £156,500 in the local group with patients concerned about the lack of postoperative follow-up arrangements on return and clinicians about the high complication rate, which may have contributed to the discontinuation of the scheme.

3.1.6 UK NHS Policy Response to EU Legislation and Internal Demand

Since the new EU Directive on cross-border health care was passed in 2011 establishing the right to receive medical care in other EEA countries, the UK DH has developed policy guidelines for patients wishing to receive planned treatments in other EU countries. The UK has European-wide agreements, where patients can obtain funding for treatment inside the EU and regulations that allow NHS purchasers to refer patients outside the EEA, although there are restrictions on patients receiving treatment outside of the EU based on the seriousness of their condition and the suitability of the treatment (NHS Choices 2013).

There are three main routes to receiving treatment in Europe and possibly elsewhere:

(1) The S2/E112 route entitles UK resident patients to treatment in another EEA country under the same terms and conditions as residents in the host country; hence any co-payments, administrative charges and waiting times need to be incurred by the patient. The NHS would then pay the amount that the host country's third party payer would reimburse the provider for the treatment. It may be possible for any co-payments and other charges to be claimed back by the patient from the NHS on returning from abroad. The S2 route can be used for more expensive treatments, including ones not available on the NHS. To go abroad for treatment to an EEA member under the S2 route requires written permission from the patient's local health commissioner (PCT or CCG). Approval depends on:

- A UK NHS consultant recommendation following a full clinical assessment that the treatment is appropriate to the patient's specific needs.
- The costs of sending the patient abroad for treatment are justified i.e. the treatment is efficient, fair and in the interest of all the needs of all patients in the locality.
- The treatment is available under the other EEA country's state health scheme.
- The patient is entitled to treatment under the NHS.

(2) An alternative route is for UK residents to access NHS funded care in Europe under Article 64 (or Article 56) of The EU Directive on cross-border health care, which grants the patient the fundamental right to access health care services in another EEA country as long as the treatment is medically necessary and is also available under the NHS. It covers treatment in both public and private hospitals. In this case, the patient will first pay the costs of treatment abroad and then claim reimbursement from the NHS on their return. However, prior authorization should be sought by the patient. The patient will be reimbursed the amount the treatment would have cost under the NHS and the patient will have to pay the difference out of their own pocket. Households entitled to NHS transport cost assistance in the UK will receive travel cost assistance to go abroad, but no assistance is available for accommodation or living costs.
(3) It may also be possible for a local NHS commissioner to approve reimbursement for treatment anywhere overseas following an Individual Funding Request. The overseas host country is not obliged to accept patients for treatment, but they will have to explain any decision to refuse treatment. Nor can patients expect to receive preferential treatment to a patient already within the country's state health care system. Unplanned health care usage while abroad is covered either by the European Health Insurance Card (for EU citizens in EEA countries), private insurance and out-of-pocket expenses.

The EU legislation implies the NHS must rule in favor of applications demonstrating "undue delay" where the equivalent treatment on the NHS is not available within a medically acceptable period. Decisions on undue delay must be based on a medical assessment and are reviewed while the patient is waiting for treatment. Waiting times given the patient's medical needs, the evidence that exists for the patient's condition (such as NICE guidance), and the likely course of the patient's medical condition are also considered.

3.1.7 The Costs of Cross Border Health Care to the UK

In total the UK paid out £561 million pounds (van Ginneken and Busse 2011) in 2006 to other EU countries for cross-border health care of which over 85 % was for the cost of state pensioners living predominantly in Ireland, Spain and France. By 2012 this had risen to nearly £900 million (Donnelly 2013). It is likely given the health status and expected expenditures of these pensioners the UK government has been overpaying for their expected costs of care (Declan 2012). The remainder of the outward costs are comprised of frontier workers temporarily resident abroad, early retired, temporary visitor expenses under EHIC and planned treatment in other EU countries. In terms of inward revenue the UK receives relatively little (£31 million) with over £163 million of unpaid debts due to treatment in the UK, while the UK owed £45 million to other EU countries in 2005 (van Ginneken and Busse 2011), which increased slightly to £50 million in 2012.

Hence, the UK runs a significant health care payments deficit with other EU countries. This appears to be due the differential number of ex-pat residents who have retired in other EU countries for which the UK government contributes to the cost of health care in their host country (E121 and E126 payments). There is also a problem with collecting accurate data on non UK residents accessing care in the UK.

3.1.8 Opening up the NHS Care Market to Private and External Foreign Providers

The latest NHS reforms passed into law under the new Health and Social Care Act (2012) have created an open market structure such that competition law applies to virtually the entirety of the NHS (High Court 2011). CCG hold competitive tenders for services from both public and private sector providers. Quasi-autonomous not for profit NHS FT are competing for services with specialized ISTC and larger private sector providers. These contracts fall within domestic, European, WTO and open market access competition law. Hence other European and international health care providers cannot be prevented from accessing the UK health care market. Section 75 regulations of the Health and Social Care Act require large sections of the NHS to be tendered to the open market and have come into force.

This is likely to impact the purchasing of any NHS goods and services if the value of the contract exceeds £156,442 (€178,750; $238,000), including clinical as well as non-clinical services. There appears little the government can do if it wished to protect the NHS from the application of competition law, other than completely removing the present functions and altering the market structure of the NHS. The promotion of patient choice has been enhanced by the policy of Any Qualified Provider (AQP), which creates competition for service contracts and patients by providers. The intention of the policy is to enable referred patients to be able to choose from a list of qualified providers from the NHS, private or charitable sector who have met NHS service quality requirements, prices and normal contractual obligations. The policy has been phased in for a number of services in primary and secondary care since April 2012. Under AQP, purchasers (now CCG) have been encouraged to extend the set of services and providers locally to include musculoskeletal services for back and neck pain; adult hearing aid services in the community; diabetes clinics, continence services; diagnostic tests closer to home; wheelchair services; primary care psychological therapies for adults and children; podiatry services; venous leg ulcer and wound healing. From October 2012, 39 more community and mental health services will be added (British Medical Association 2012). Purchasers were obliged under the procurement regime to conduct open competitions before appointing or commissioning health care providers. Hence, instead of patients in poorly served areas having to seek out and travel further for treatment from alternative providers at home or abroad, newly autonomous NHS as well as private sector providers from the UK or abroad can enter local markets to meet patients' needs.

Of the 87 providers running services under AQP worth £262 m ($399 m; €300) in England, 38 are from the private sector, 26 are within the NHS, 18 are charities, four are social enterprises, and one is a voluntary organization. In the first wave, 398 contracts were signed in eight NHS areas. The NHS Support Federation found that contracts for around 100 NHS clinical services totaling almost £1.5 billion (€1.7 billion; $2.2 billion) have since been advertised from 1 April 2013, with private companies winning three quarters of the contracts (Iacobucci 2013).

Contracting with the private sector has already occurred for a number of elective surgery services with ISTC providing highly focused clinics and surgical interventions in specialties such as orthopedics and have attracted significant market share from NHS hospitals in certain specialties (Arora 2013). Examples of foreign multinational organizations entering this market include South Africa's Netcare, Cinven/Spire (European holding), Apollo of India, and the Swedish Capio Group (a leading private health care company in Europe) who later sold most of its UK hospitals to Ramsay Health Care (an Australian company) in 2008 (Lethbridge 2011).

The market for community and preventive health care services has attracted some of the large US managed health care organization such as UnitedHealth with their Evercare specialized community nurses program as well as HealthDialog, which merged with BUPA to supply predictive patient risk services, demand management and commissioning services to CCGs (Lethbridge 2011). The provision of laboratory services at the Royal Free and UCL Foundation Trust has been awarded to the Australian company Sonic health care, a global provider of clinical diagnostics with a turnover of 3 billion $Aus (£1.78 billion) via its UK subsidiary The Doctors Laboratory who are tendering for contracts in the NHS pathology sector (Foot 2013). The increase in immigrants from Eastern Europe has also created a market for private health care providers that meet the needs of the Polish community in the UK who are willing to pay for what they perceive to better treatment than NHS services. Since 2007, private Polish medical centres have sprung up all over the UK with at least 20 in London, and others in Manchester, Reading, Bristol and Glasgow (Williams 2013).

3.2 *Patient and Health Care Mobility into the UK NHS from Importing Countries*

The UK has always attracted patients from abroad as private patients who have come specifically to see specialists either privately in their clinic at NHS hospitals or in private hospitals. Private patients are likely to pay out of pocket or have international health insurance or be exempt from charges under overseas agreements from EU cross-border rights or outsourcing government agreements with the UK.

Overseas visitors from abroad who are not 'ordinarily resident' in the UK are entitled to be treated in the NHS and, depending on the nature of treatment, may or may not be charged. Under current regulations, overseas visitors to Britain can in principle register with a GP practice (although some GP practices are now demanding proof of a UK address), obtain an NHS number and be seen by a GP free of charge. They can also be referred to a secondary care specialist for NHS treatment in an outpatient (ambulatory) or hospital inpatient setting. However, legally, they will incur overseas visitor charges unless the treatment received is listed as an exempted condition. Urgent treatment in an accident and emergency department (A&E) does not incur any overseas visitor charges, although subsequent inpatient admission is chargeable. Family planning and infectious disease diagnosis and acute treatment are also exempted. Only the costs of inpatient hospital and specialist outpatient visits are chargeable for the full cost of treatment under national fixed rate PbR tariffs, excluding VAT.

There has been a steady flow of 'international patients' traveling to the UK to access a range of health care services with estimates of 52,000 overseas visitors to the UK for medical treatment spending £132 million in 2010 (Lunt et al. 2011; Office for National Statistics (ONS) 2011). Williams et al. (2000) suggested that, over a decade ago, there were 16,600 admissions from overseas into the UK private health care sector. The remainder of overseas patients will be seen privately in the NHS. The number of visitors from abroad privately consulting NHS specialists in NHS hospitals in England liable to be charged or covered by cross-border agreements is difficult to identify precisely from Hospital Episode Statistics data (HSCIC 2013). However, it appears to be somewhere between 11,000 and 23,000 patients during the financial year 2010. Older data relating to the period between 2003 and 2005 found the number of EU patients with planned treatment either as an outpatient or inpatient (under E112 forms) in UK hospitals increased between 2000 and 2005 from 747 and 976 patients with the majority from Ireland and Italy (van Ginneken and Busse 2011).

To access specialist NHS treatment without incurring private patient charges it is necessary to be "ordinarily resident" in the UK. Under current guidelines, it is the responsibility of the NHS Hospital Trust to determine the "ordinarily resident" status of the patient and liability for charges for NHS treatment. Ordinary residence is not, however, well defined in the regulations, and there is no minimum period of residence that confers ordinarily resident status. It broadly implies "living lawfully in the United Kingdom voluntarily and for settled purposes as part of the regular order of their life for the time being, whether they have an identifiable purpose for their residence here and whether that purpose has a sufficient degree of continuity to be properly described as "settled". It is complex to adjudicate as it is based on a fact from a previous legal case, and there appears to be considerable confusion about the current regulations within the NHS. Other criteria can also determine eligibility for residency status for NHS treatment. For instance, if someone has a work contract with a UK based company that entitles them to be resident in the UK; they have been resident in the UK for the previous year, or for more than 10 years prior to being abroad for more than a year and less than 5 years. If the patient has been resident in the UK or is a UK pensioner spending at least 182 days or more in the 12 months prior to treatment, and has their 'centre of interests' in the UK (for instance property, family, a work contract or a place as a full time student on a course of at least 6 months) (Department of Health 2012). It should be noted that NHS hospitals may not have much incentive to identify overseas visitors as they will be guaranteed reimbursement if the patients are categorized as NHS patients, whereas they risk non-payment from overseas visitors and substantial administration costs in dealing with invoices, claims and chasing payments from overseas visitors.

Recently, there has been growing political and public interest in the potential abuse of the system by overseas visitors from outside the EEA deliberately seeking to obtain free treatment on the NHS. In a statement about immigration, the current UK Prime Minister David Cameron said "What we have is a free national health service, not a free international health service" (Chalabi 2013). In the latest 2012 survey of 'Public Perceptions of the NHS and Social Care' undertaken by IPSOS/MORI, 5 % of the public surveyed believed immigration was one of the biggest problems facing the NHS. Analysis of data from a sample of NHS bodies conducted by the DH in England suggests around £33 million of treatment charges were issued to overseas visitors of which 11.5 million (35 %) were unpaid with around 4,500 people with outstanding NHS debts of over £500. Given that the NHS is being asked to find 'efficiency savings' of £20 billion of its £109 billion budget, eliminating fraudulent 'health tourism' equates to fulfilling 0.058 % of the savings. A DH analysis suggested that fewer than half of overseas visitors using hospitals are identified and only half their costs recovered. Some unofficial reports have speculated the cost is higher. CCI insurers (2003) estimated it at £50 million–£200 million (Boffey 2013). Others believe the figure should be higher with the true cost in billions rather than millions because health tourists should be charged private health care prices rather than the lower charges which the NHS system offers. However, this claim seems unfounded given that patients may have only fallen ill while visiting the UK, and accessed care through the NHS, incurring normal waiting times and standard NHS wards and not the preferential treatment offered to private patients. The UK Border Agency has analyzed passenger arrivals and found evidence that some visitors were accessing health care and often sought to evade payment on a repeat basis. From October 2011 the immigration rule has been amended to allow the refusal of a new visa or extension of stay to a person with an outstanding debt to the NHS of £1,000 or more (Home Office Borders Agency 2013).

The coalition government announced it would undertake a fundamental review of the rules and practices around determining eligibility and charging non EU overseas visitors for NHS treatment with the view that the current criteria and arrangements for regulating entitlement to free NHS care is not efficient or fair compared with policies in other countries. The review will consider qualifying residency criteria for free treatment; charges for GP or other services outside of hospital; improvements in screening for eligibility and recovery of charges; and the introduction of health insurance tied to visas (Powell 2012). This has resulted in proposed plans for new migrants from outside the EEA to pay an additional £200 per year, and up to £1,000 over 5 years on top of the average £500 visa application fees to cover the cost of their health care. More than 270,000 migrants, who come to study and work in Britain from outside the EU each year, are likely to be affected by the levy. The proposal also wants to introduce charges for GP visits by overseas visitors and for GPs to take responsibility for determining the residency status of patients and liability for NHS charges (Travis 2013). Estimates of the combined cost of EU and non-EU migrants to the NHS is only £7 million (Williams 2013); their cost is far smaller than their number, proportionally, because they are younger, on average, than indigenous Britons; they are less likely to be ill and they are less likely to have started a family. Wadsworth (2012) found the distributions of self-assessed health and use of hospital services to be similar between migrants and non-migrants in the UK with migrants using GP services slightly more frequently.

Health professionals have raised concerns about introducing user charges for accessing primary care and the burden of assuming responsibility for "immigration control". There are public health concerns if visitors to the UK are reluctant to visit their GP, leading to delayed diagnosis and treatment of infectious diseases contributing to their spread. It may also cost the NHS more if patients receive more expensive treatment in A&E (Boffey 2013). Politicians have objected to the idea as it signals that Britain is not an open society, and may dissuade foreign workers (many of whom work in the NHS) and students from coming.

3.2.1 Exporting NHS Health Care into Foreign Countries

The NHS is seeking to develop its export markets for health care services. Under the 2012 Health and Social Care Bill, the income cap for NHS Trusts on private patient earnings has been relaxed from 2002/3 levels to 49 %. This provides FT with international reputations (such as Great Ormond Street) the possibility to attract more international private patients, which was specifically mentioned in the new reforms. There needs, however, to be a justification for how non-NHS income will benefit NHS services. Private UK health care companies, notably BUPA have already moved into the international market (Lethbridge 2011) and NHS Trusts are beginning to enter into contracts to provide services abroad. Moorfields Eye Hospital in London has already set up a hospital in Dubai, and Great Ormond Street Children's Hospital provides services in the Middle East (Boseley 2012). Hospitals are to be encouraged by the government to sell their services abroad by setting up clinics or hospitals with the worldwide reputation of the NHS brand in order to generate overseas revenue (although the NHS brand may have suffered with the negative publicity surrounding recent poor care scandals). The UK government has set up a body called Health Care UK to assist the NHS in attracting contracts overseas, which has been put together by the DH and the UK Trade and Investment department (UKTI) with potentially lucrative contracts in growing markets such as India and China. The board is said to be recruiting a managing director at a salary of £100,000.

4 Conclusions

Health care is characterised by significant uncertainty, asymmetric information and risks from externalities. These factors have implications for patient mobility because health care markets may not function as intended, and patient's preferences are not always met. Illness is uncertain and sudden health shocks can leave individuals immobile and unable to make rational choices. Externality risk, such as the exposure to the effects of air pollution, infectious diseases and resistant microbes may increase with mobility. Patients value the agency relationship of their GP or specialist and often delegate responsibility for health decisions.

Nevertheless, patients appear to value greater choice, information and shared decision making. They are willing to travel further to obtain faster access, better value, quality and innovative care. Patients in the UK rely on a range of information, but seem to prefer experience, the knowledge of others or informal information rather than statistical analysis or reported comparative outcome measures. They are becoming more mobile with an increase in the frequency and distance of their medical visits, which may reflect rising population health needs, a greater propensity to seek health care as well as improvements in the supply of health care services such as shorter waiting times. The effectiveness of any policy to promote patient choice and mobility depends on the ability of patients to identify the most efficient providers and to travel beyond their local providers if necessary. The English Patient Choice initiative may have increased mobility and altered patient flows in response to local problems with accessibility and quality of care. There is evidence that patients' choices are associated with objective measures of quality, and that patients' sensitivity to quality in their choice of hospital increased (Gaynor et al. 2012). Hence, analysing changes in patients' choices of treatment provider as well as market shares of providers may reveal relative differences in health care performance and will continue to be an important area of research.

In future, however, the need to travel to access health care may diminish. There is growing use of technology such as home telemedicine services and the internet to provide consultation with medical practitioners online. Personal medical devices enable patients to take responsibility for the management of their health condition and share data remotely with medical practitioners. Pharmaceuticals or other health care products can be prescribed and purchased on the internet with mail order pharmacies delivering prescriptions. Service delivery may also be changing with an increasing home care service sector, community nurses visit patients as personal case managers, and long term care residents receive significant levels of health care where they live.

Increased patient choice and market competition risk undermining the equity principles of the NHS of ensuring equal health care availability for equivalent health care needs. Inequality in health care use and subsequently health may be exacerbated by choice as the better educated or economically advantaged are more health literate, better informed and mobile. They can, therefore, afford the increased search and travel costs, obtaining faster access to the best services. There is also a view that greater choice would be good for health care inequalities. Studies have indicated that the poor, the elderly, individuals from ethnic backgrounds and those with previous poor treatment valued choice more (Dixon and Le Grand 2006; Robertson and Burge 2011). These are likely to be groups disadvantaged in the health care system who face worse local services, access problems, and discrimination. Hence, the opportunity to have more autonomy, and access to better services elsewhere would be of more value. There were already significant socio-economic inequalities in the utilization of specialist health care, uptake of preventive care and waiting times prior to the choice reforms (Dixon et al. 2007). Evidence to date is limited, but trends for selected indicators of health care inequality show that patient choice did not worsen socio-economic related inequality (Cooper et al. 2009; Cookson et al. 2013). The patient choice surveys implied that access inequalities may be a consequence of the choice policy with only 50 % of the population offered choice with differences across demographic, ethnic and socio-economic groups. There was also a social gradient in individuals willing to travel further (Robertson and Burge 2011). Individuals offered choice were more likely to obtain appointments at a preferred provider, be seen on a more convenient day and time and have shorter waiting times. Dusheiko and Gravelle (2012) found that GP practices with higher rates of Choose and Book utilization (electronic booking system used to facilitate patient choice) had significantly lower rates of patient non-attendance at first outpatient appointment. This would suggest that there is the possibility that choice could widen health inequalities. This is another important area for future research.

On the supply side, greater competition for patients, the PbR fixed national tariff, opening the market to more competitors and shifting more services into the community from the hospital sector may have been of benefit to patients, giving them more choice and local options, reducing travel costs and potentially waiting times. It may also have lead health care providers to improve quality in order to retain and attract new patients as well as become more efficient providers, reducing length of stays. Opening up the market to more competition nationally and internationally can reduce the market power of NHS clinicians who had an incentive to maintain longer waiting lists to encourage patients to seek treatment privately with them.

Competition, however, could exacerbate the variability in quality of care with providers losing market share, revenue, and having to cut back on services and resources. This could leave individuals dependent on local services worse off and put at risk the quality of emergency or other essential services, which could leave the immobile and most vulnerable worse off. The entry of foreign and private providers is perceived as a threat to the stability of the public NHS and there is opposition to increasing open market competition for more NHS services. It risks creating inefficient duplication of services leading to over capacity and fragmented care. Hence, the NHS could lose some of its advantages from economies of scale and scope as well as opportunities to improve care co-ordination. There is evidence that private providers tend to attract healthier patients and burden public hospitals with more difficult cases, but also that ISTC are more specialized and provide efficient treatment (Siciliani et al. 2013). Furthermore a reduction in patient volumes could reduce skill levels in public hospitals. Private companies can decide to exit the market when they choose, leaving areas with a void in service delivery forcing patients to travel.

A high proportion of patients would be willing to travel for improved health outcomes. Hence, one would expect higher rates of patient mobility abroad, especially given the evidence of fairly average patient outcomes in cancer, and across a number of quality dimensions. One reason for the low willingness to travel is financial. The UK NHS has very few financial barriers to accessing care with only 5 % of individuals having reported any financial barriers (Commonwealth Fund 2011). User satisfaction for quality of the health service was also high at over 85 %, which was above the OECD average. In the last decade there has been significant additional investment in the health care service. Amenable mortality rates in the UK between 1997 and 2007 decreased by between 4.6 and 5.2 % one of the largest reductions in the OECD. Expectations, trust of statistics and experience also appeared important with many patients uncertain about the quality of care offered elsewhere. There is patient loyalty to their local health providers and a desire to retain specialist services close to home is illustrated by the recent campaign to retain pediatric heart surgery at Leeds General Infirmary despite concerns raised about the quality of care at the hospital (Boseley 2013a, b).

The new EU directives have removed some of the financial barriers to mobility, but utilization rates for the scheme still seem low. This may reflect the high search and transaction costs as well as administrative barriers in place that may restrict access and uptake of the scheme. Patients will also need to be able to navigate the system of their chosen host country. International patient mobility was not wholly supported by UK medical practitioners who raised clinical, safety, regulatory and economic concerns. There are significant clinical challenges in trying to co-ordinate aftercare when patients return from overseas and it imposes additional demands on the NHS, particularly if there are subsequent complications following treatment. There is a potential role for UK physicians to assist with patient mobility abroad, especially if there are limited options domestically, but in reality it is unclear they would have the means and time to support patients. Competition from foreign providers and reductions in waiting times would also threaten their income from private practice as well as their influence on the local health economy; hence, the incentives for domestic providers to support patient mobility may be weak. Even the NHS Patient Choice initiative received a mixed response from physicians in particular GPs. There was a discrepancy between the opinion of GPs and their patients about the value of choice and significant variation in offering choice to patients among certain GPs who may have perceived it as a challenge to their clinical responsibility and judgement. They may also have wanted to protect their local health economy, preserving existing referral networks and established professional relationships with local clinicians. Hence, discouraging choice and mobility may have been a strategic response based on professional self-interest and not patient altruism.

Policy makers had a more open minded view on the opportunities created by the EU legislation facilitating patient mobility and trade in health services. They were quick to pilot initiatives to outsource treatment to neighboring EU countries given significant domestic waiting times, patient and political pressure. The pilots demonstrated that sufficient domestic demand existed, and the feasibility of identifying high quality overseas providers with the capacity, incentives and willingness to treat UK patients. The pilots were successful from the perspective of the patients who received faster and good clinical care. The volume of patients actually sent was low and average costs of treating them were quite high, between that of private UK providers and UK NHS treatment costs. Whether or not this additional cost was justified by the reductions in patient waiting times, earlier improvement in health status as well as the opportunity cost of spending elsewhere in the local NHS economy was not fully evaluated.

After 4 years of pilots to treat patients overseas the formal use of foreign providers was largely discontinued. The relative costs and benefits of sending patients overseas were no longer favorable. Increased investment in the NHS, greater use of additional capacity in the domestic private sector and generous contracts with ISTCs as well as NHS specialist consultant contracts combined with strong managerial incentives significantly reduced patient waiting times. The later implementation of patient choice policies accompanied by PbR activity based reimbursement also facilitated internal movement of patients from areas with high waiting time providers and promoted the use of alternative providers able to maintain shorter waiting times. Hence, instead of exporting patients in poorly served areas abroad, patients could seek treatment from alternative providers at home by being prepared to choose and travel beyond local hospitals.

The mass outsourcing of patients abroad did not appear a viable long term solution for the UK. However, there is a policy for importing patient care from abroad on an ad hoc basis with requests for funding approval left to the discretion of local health care purchasers. Information is clearly available on the internet, and a dedicated international patient advice service with a telephone helpline is available, but the onus is on the patient to research, organise and apply for funding from their local purchasing group (CCG). There were fears that the new EU directives could undermine policies to ensure cost-effective use of new technologies in particular pharmaceuticals in the UK through NICE reimbursement guidance. NICE has been criticized for the time it takes to evaluate new treatments which may delay innovative technologies. There is also variability in local implementation of NICE guidance. Hence, patients may be able use the "undue delay" ruling to by-pass NICE decision making and obtain earlier treatment in EU countries (and possibly anywhere overseas) that have already given market authorization for treatments still under review by NICE. If NICE subsequently advised against NHS reimbursement for a treatment already approved for use in another EU country this could result in refusal by the local purchasing organization to refund the treatment cost. To date, however, patient mobility abroad has not proven a significant enough phenomenon to destabilise the local health economy. It has, however, generated considerable policy interest as well as the attention of service providers, which has lead to numerous policy initiatives.

Attracting private patients from abroad could be an important source of revenue. NHS patients may benefit if new investment leads to enhanced facilities that are shared between private and NHS patients. Delivering care to international patients contributes to the level of direct foreign exchange earnings coming into the UK and sectors other than medical care especially those associated with hospitality and travel may also benefit financially from increased medical tourism, which could create employment and raise GNP. Central government will also benefit through increased taxation revenue. The increased ability to purchase the latest technology, for example, and treating foreign patients may broaden the case-mix for staff, or increase throughput to enable them to become more skilled and deliver higher quality care. Health trade and the need to ensure the NHS brand is internationally recognized as a provider of high quality, efficient care could exert competitive pressure on systems which are both exporting and importing health care, which could have implications for both the costs and quality of service delivery depending on the nature of the competition. There is, however, a lack of evidence regarding the economic benefits and costs of incoming medical tourists, and little information on how the profits from private practice have benefitted NHS patients.

The search for income and profits from private and international markets may be detrimental to domestic patients. There is a risk that private profits will be invested into more private sector projects or be paid in rents to clinicians or managers (Harris 2012). Many of the private health care providers in the UK are overseas subsidiaries; hence much of the profit may leave the UK. The opportunity cost of the resources devoted to treating private patients from overseas at the expense of treating NHS patients may be high. It could lead to a growth in waiting lists and waiting times for NHS patients or underinvestment in important A&E services.

In general it does appear that mobility can enable patients to obtain better care experience, outcomes and value for money. In doing so, they have to take on significantly more personal responsibility and risk, but when properly informed and researched as well as supported by a professional structure the risk can be minimized. The resources for facilitating patient mobility in the UK still seem limited, which in part is due to scepticism and resistance from some members of medical profession in the UK. There also appears to be a need for greater regulation, patient protection and co-ordination of care for patients traveling abroad. The volume of patients exercising choice and traveling abroad for treatment appears to be less than anticipated. Nevertheless the numbers have been growing significantly enough to have impacted on the market and organization of health care provision in the UK.

Acknowledgments I am grateful for the support of the UK Medical Research Council (MRC) Early Career Fellowship Grant in the Economics of Health (Reference Number G0902067). I would also like to thank the Editors of the book for their invitation, encouragement and advice as well as Doriana Delfino and Adrien Magnenat for their invaluable help with preparing the chapter. All subsequent mistakes are the responsibility of the author alone.

References

Appleby, J. (2012). British Social Attitudes Survey 2012: Public satisfaction with the NHS and its services. The King's Fund. http://www.kingsfund.org.uk/projects/bsa-survey-2012. Accessed 29 July 2013.

Appleby, J. (2013). Spending on health and social care over the next 50 years: Why think long term? Kings Fund.

Arora, S., et al. (2013). Public payment and private provision. Nuffield Trust.

Beecham, L. (2002). British patients willing to travel abroad for treatment. *British Medical Journal, 325*(7354), 10.

Bertinato, L., et al. (2005). *Cross-border health care in Europe*. Copenhagen: European Observatory on Health Systems and Policies.

Bevan, G., & Hood, H. (2006). Have targets improved performance in the English NHS? *British Medical Journal, 332*(7538), 419–422.

Birch, J., et al. (2007). The complications of 'cosmetic tourism': An avoidable burden on the NHS. *Journal of Plastic Reconstructive and Aesthetic Surgery, 60*(9), 1075–1077.

Björnberg, A. (2012). Health Consumer Powerhouse Euro Health Consumer Index 2012 report. Health Power House. http://www.healthpowerhouse.com/files/Report-EHCI-2012.pdf. Accessed 29 July 2013.

Boffey, D. (2013). GPs told to charge non-Britons for NHS services. The observer (Guardian online). http://www.guardian.co.uk/uk/2013/jun/29/jeremy-hunt-immigrants-free-nhs-services. Accessed 29 July 2013.

Boseley, S.(2012). NHS 'brand' could be sold abroad to generate income, says government. The Guardian. http://www.guardian.co.uk/society/2012/aug/21/nhs-brand-sold-overseas-hospitals. Accessed 29 July 2013.

Boseley, S. (2013a). Cosmetic surgery tourism on rise as satisfied customers spread the word. The Guardian. http://www.guardian.co.uk/lifeandstyle/2013/jun/21/cosmetic-surgery-tourism. Accessed 29 July 2013.

Boseley, S. (2013b). Leeds child heart surgery turf war is new low. The Guardian.

Boseley, S. (2007). Plastic surgeons attack botched holiday surgery. The Guardian. http://www.theguardian.com/society/2007/nov/19/health1?INTCMP=SRCH. Accessed 30 July 2013.

Boyd, J. (2007). The 2006 inpatients importance study. The Picker Institute. http://www.nhssurveys.org.

Bridgewater, B. (2013). Publishing cardiac surgery mortality rates: Lessons for other specialties. *British Medical Journal, 346*, f1139.

British Medical Association (2012). What we know so far…Choice and any qualified provider. Health Policy and Economic Research Unit.

Busse, R., et al. (2011). Access to health care services within and between countries of the European Union. In M. Wismar, W. Palm, J. Figueras, K. Ernst, & E. van Ginneken (Eds.), *Cross-border health care: Mapping and analysing health systems diversity* . Copenhagen: The European Observatory on Health Systems and Policies.

Cantarero, D. (2006). Health care and patients' migration across Spanish regions. *European Journal of Health Economics, 7*, 114–116.

Chalabi, M. (2013). Health tourism: How much does it cost the NHS? The Guardian online. http://www.guardian.co.uk/politics/reality-check/2013/jul/03/health-tourism-cost-nhs-jeremy-hunt. Accessed 29 July 2013.

Chilingerian, J., et al. (2012) Globalization and health. In L. R.Burns, E. H. Bradley, & B. J. Weiner (Eds.), *Shortell & Kaluzny's health care management* (6th ed.). London: Delmar Cengage Learning.

Cooper, Z., et al. (2009) Equity, waiting times, and NHS reforms: retrospective study. *British Medical Journal, 339*. doi: 10.1136/bmj.b3264. Accessed 29 July 2013.

Cookson, R., et al. (2013). Does hospital competition harm equity? Evidence from the English National Health Service. *Journal of Health Economics, 32*, 410–422.

Commonwealth Fund (2011). Why not the best? Results from the US scorecard on health system performance. Commonwealth Fund. http://www.commonwealthfund.org/Publications/Fund-Reports/2011/Oct/Why-Not-the-Best-2011.aspx?page=all. Accessed 29 July 2013.

Culley, L., et al. (2011). Crossing borders for fertility treatment: Motivations, destinations and outcomes of UK fertility travellers. *Human Reproduction, 26*, 2373–2381. doi:10.1093/humrep/der191.

Dawson, D., et al. (2007). The effects of expanding patient choice of provider on waiting times: Evidence from a policy experiment. *Health Economics, 16*(2), 113–128.

Declan, L. (2012). BBC Panorama: Britain's Secret Health Tourists. http://chr1sr0berts.wordpress.com/2012/11/12/bbc-coverage-of-the-nhs-panorama-britains-secret-health-tourists-a-case-study/. Accessed 29 July 2013.

Delamothe, T. (2008a). NHS at 60: Founding principles. *British Medical Journal, 336*, 1216–1218. doi: http://dx.doi.org/10.1136/bmj.39582.501192.94. Accessed 29 July 2013.

Delamothe, T. (2008b). NHS at 60: A comprehensive service. *British Medical Journal, 336*, 1344–1345. doi: http://dx.doi.org/10.1136/bmj.39582.501192.94. Accessed 29 July 2013.

Deloitte (2009). *Medical tourism: Update and implications*. Washington, DC: Deloitte Centre for Health Solutions. http://www.deloitte.com/assets/Dcom-UnitedStates/Local%20assets/documents/us_chs_medicaltourism_111209_web.pdf.

Department for Transport. (2004). Travel Statistics Bulletin: National Transport Statistics (NTS) 2002.

Department of Health. (2002). *Reforming NHS financial flows: Introducing payment by results*. London: Department of Health.

Department of Health. (2005). *The NHS Improvement Plan: Putting people at the heart of public services*. London: Department of Health.

Department of Health. (2008a). Delivering the NHS Plan. HMSO.

Department of Health. (2008b). High quality care for all: NHS Next Stage Review (final report).

Department of Health. (2010). Equity and excellence: Liberating the NHS. Department of Health.

Department of Health. (2012). Guidance on implementing the overseas visitors hospital charging regulations.

DHSS. (1989). *Working for patients*. London: HMSO.

Dixon, A., & Le Grand, J. (2006). Is greater patient choice consistent with equity? The case of the English NHS. *Journal of Health Service Research Policy, 11*, 162–166.

Dixon, A., et al. (2007). Is the British National Health Service equitable? The evidence on socioeconomic differences in utilization. *Journal of Health Services Research and Policy, 12*, 104–109.

Dixon, A., et al. (2010). Patient choice: How patients choose and how providers respond. Findings from the National Patient Choice Survey. The King's Fund. http://www.kingsfund.org.uk/projects/patient-choice-how-patients-choose-and-how-providers-respond. Accessed 29 July 2013.

Dixon, J., et al. (2011). A person based formula for allocating commissioning funds to general practices in England: Development of a statistical model. *British Medical Journal*. doi:10.1136/bmj.d6608.

Dixon, P., et al. (1997). *Patient movements and patient choice: Report for National Health Service Executive*. London: York Health Economics Consortium.

Dixon, S. (2010). Report on the National Patient Choice Survey. Department of Health. https://www.gov.uk/government/uploads/system/uploads/attachment_data/file/216073/dh_117096.pdf. Accessed 29 July 2013.

Donnelly, L. (2013). NHS counts £900 million cost of treating EU visitors. The Telegraph 2013. http://www.telegraph.co.uk/health/healthnews/10022601/NHS-counts-900-million-cost-of-treating-EU-visitors.html. Accessed 29 July 2013.

Dr Foster Intelligence (2012) Fit for the future? Dr Foster Hospital Guide. http://download.drfosterintelligence.co.uk/Hospital_Guide_2012.pdf. Accessed 29 July 2013.

Drösler, S., et al. (2009). OECD health working papers no. 47. Health Care Quality Indicators Project: Patient Safety Indicators Report 2009. http://www.oecd.org/health/health-systems/44192992.pdf. Accessed 29 July 2013.

Drummond, M., & Sorenson, C. (2009). Nasty or nice? A perspective on the use of health technology assessment in the United Kingdom. *Value in Health, 12*(2), 8–13. doi:10.1111/j.1524-4733.2009.00552.x.

Dusheiko, M., et al. (2004). The effect of practice budgets on patient waiting times allowing for selection bias. *Health Economics, 13*(10), 941–958. doi:10.1016/j.jhealeco.2005.08.001.

Dusheiko, M., et al. (2008). Explaining trends in concentration of health care commissioning in the English NHS. *Health Economics, 17*(8), 907–926.

Dusheiko M., & Gravelle, H. (2012). The impact of choose and book on outpatient appointment non-attendances. In *CES-HESG conference*, Aix en Provence, January. http://www.ces-asso.org/autres-conferences. Accessed 29 July 2013.

European Commission (2006). Health & Consumer Protection Directorate-General: Report on the work of the High Level Group in 2006. European Commission. http://ec.europa.eu/health/ph_overview/co_operation/mobility/docs/highlevel_2006_007_en.pdf. Accessed 29 July 2013.

Evans, R. G. (1984). *Strained mercy: The economics of Canadian health care*. Toronto: Butterworths.

Ferguson, N. (2003) British imperialism revisited: The costs and benefits of "Anglobalization". *Historically Speaking, 4*(4), 21–27. Accessed 29 July 2013.

Foot, T. (2013). As blood tests are switched to 'joint venture' with company, NHS staff are told: 'Join private firm or quit'. *The Camden New Journal*. http://www.camdennewjournal.com/news/2013. Accessed 29 July 2013.

Francis, R. (2013). The Mid Staffordshire NHS Foundation Trust Public Inquiry. Report of the Mid Staffordshire NHS Foundation Trust Public Inquiry: Executive summary. HMSO. http://www.midstaffspublicinquiry.com/. Accessed 29 July 2013.

Gay, J. G., et al. (2011). Mortality amenable to health care in 31 OECD countries: estimates and methodological issues. *OECD Health Working Papers*,. doi:10.1787/5kgj35f9f8s2-en.

Gaynor, M. (2012). Free to choose? Reform and demand response in the British National NBER working paper. http://www.nber.org/papers/w18574. Accessed on 31 July 2013.

Geddes, C. C., et al. (2008). Outcome of patients from the west of Scotland traveling to Pakistan for living donor kidney transplants. *Transplantation, 86*, 1143–1145. doi:10.1097/TP.0b013e318187cdd0.

Glinos, I. A., et al. (2006). Cross-border contracted care in Belgian hospitals. In M. Rosenmöller, M. McKee, & R. Baeten (Eds.), *Mobility in the European Union learning from experience*. http://www.euro.who.int/__data/assets/pdf_file/0011/120332/E88697.pdf. Accessed 29 July 2013.

Glinos, I. A., et al. (2010). A typology of cross-border patient mobility. *Health & Place, 16*(6), 1145–1155.

Hall, C. M., & James, M. (2011). Medical tourism: Emerging biosecurity and nosocomial issues. *Tourism Review, 66*(1/2), 118–126.

Hanna, S. A., et al. (2009). Sending NHS patients for operations abroad: Is the holiday over? *Annals of the Royal College of Surgeons of England, 91*(2), 128–130.

Harris, E. (2012). The real scandal of the NHS is not about beds in India. The Guardian. http://www.guardian.co.uk/commentisfree/2012/aug/21/nhs-brand-private-profits. Accessed 29 July 2013.

Health care Commission, Audit Commission (2008). *Is the treatment working? Progress with the NHS system reform programme*. London: Audit Commission.

High Court of Justice Legal Advice (2011). In the matter of the health and social care bill and the application of procurement and competition law, advice. http://38degrees.3cdn.net/63442740413df6b835_clm6ib678.pdf. Accessed 29 July 2013.

Holliday, R., et al. (2013). Sun, sea, san and silicon. mapping cosmetic tourism. University of Leeds/ESRC project report. http://www.academia.edu/4019774/Sun_Sea_Sand_and_Silicone_Report_Summary. Accessed 29 July 2013.

Health and Social Care Act (2012). TSO http://www.legislation.gov.uk/ukpga/2012/7/pdfs/ukpga_20120007_en.pdf. Accessed 29 July 2013.

Home Office, UK Border Agency (2013): General grounds for refusal. http://www.ukba.homeoffice.gov.uk/sitecontent/documents/policyandlaw/modernised/general-grounds-refusing/leave-to-remain.pdf?view=Binary. Accessed 29 July 2013.

HSCIC (2013). Hospital episode statistics. http://www.hscic.gov.uk/hes. Accessed 29 July 2013.

Huynen, M., et al. (2005). The health impacts of globalisation: A conceptual framework. *Globalization and Health, 1*, 14. doi:10.1186/1744-8603-1-14.

Iacobucci, G. (2013). Majority of £1.5 bn NHS contracts go to private providers, study finds. *British Medical Journal, 347*. doi:10.1136/bmj.f4476.

Ireni-Saban, L. (2013). Give me children or else I die: The politics and policy of cross-border reproductive care. *Politics and Policy, 41*(1), 5–38. doi:10.1111/polp.12004.

IPSOS/MORI. (2007). Cancer values survery part two research study conducted for Cancer-backup. IPSOS/MORI. http://www.ipsos-mori.com/Assets/Docs/Archive/Polls/cancerbackup.pdf. Accessed 29 July 2013.

IPSOS/MORI. (2012). Public perceptions of the NHS survey. Department of Health. http://www.ipsos-mori.com/researchpublications/researcharchive/2975/Public-Perceptions-of-the-NHS.aspx. Accessed 29 July 2013.

Jeevan, R., & Armstrong, A. (2008). Cosmetic tourism and the burden on the NHS. *Journal of Plastic Reconstructive Aesthetic Surgery, 61*(12), 1423–1424.

Kelly, E, & Tetlow, G. (2012). Understanding competition and choice in the NHS. IFS/Nuffield Trust.

Kennedy, I. (2001). Learning from Bristol: The report of the public inquiry into children's heart surgery at the Bristol Royal Infirmary 1984–1995.

Kangas, B. (2010). Traveling for medical care in a global world. *Medical anthropology: Cross-cultural studies in health and illness* (Routledge) *29*(4), 344–362. doi: 10.1080/01459740.2010.501315. Accessed 29 July 2013.

Keogh, B. (2013). Review into the quality of care and treatment provided by 14 hospital trusts in England: Overview report. NHS choices. http://www.nhs.uk/NHSEngland/bruce-keogh-review/Documents/outcomes/keogh-review-final-report.pdf. Accessed 29 July 2013.

Krishnan, N., et al. (2010). Organ trafficking for live donor kidney transplantation in Indonesians resident in the West Midlands: High activity and poor outcomes. *Transplantation, 89*(12), 1455–1461. doi:10.1097/TP.0b013e3181da6019.

Kumarasamy, K. K., et al. (2010). Emergence of a new antibiotic resistance mechanism in India, Pakistan, and the UK: A molecular, biological, and epidemiological study. *The Lancet Infectious Diseases, 10*(9), 597–602. doi:10.1016/S1473-3099(10)70143-2.

Laura, S., et al. (2013). From data to decisions? Exploring how health care payers respond to the NHS Atlas of Variation in Health care in England. *Health Policy* . doi: 10.1016/j.healthpol.2013.04.014. Accessed 29 July 2013.

Leathard, A. (2000). *Health care provision: Past, present and into the 20th century* (2nd ed.). Cheltenham: Nelson Thormes.

Lethbridge, J. (2011). Understanding multinational companies in public health systems, using a competitive advantage framework. *Globalization and Health, 7*, 19. doi:10.1186/1744-8603-7-19.

Levaggi, R., & Zanola, R. (2004). Patients' migration across regions: the case of Italy. *Applied Economics*, *36*, 1751–1757.

Lowson, K., et al. (2002). *Evaluation of treating patients overseas*. York: York Health Economics Consortium and Department of Health.

Loh, C. P. (2013). Health tourism on the rise? Evidence from the balance of payments statistics. *European Journal of Health Economics,*. doi:10.1007/s10198-013-0521-0.

Lunt, N., et al. (2010). Nip, tuck and click: Medical tourism and the emergence of web-based health information. *Open Medical Informatics Journal,* 4, 1–11. http://www.ncbi.nlm.nih.gov/pmc/articles/PMC2874214/. Accessed 29 July 2013.

Lunt, N., et al. (2011). Medical tourism: Treatments, markets and health system implications: A scoping review. OECD. http://www.oecd.org/els/health-systems/48723982.pdf.

Lunt, N., et al. (2013). A framework for exploring the policy implications of UK medical tourism and international patient flows. *Social Policy and Administration, 47*(1), 1–25. doi:10.1111/j.1467-9515.2011.00833.x.

Masiero, G. (2002) Patient choice of practice in the market for primary care. PhD Thesis, York.

McKelvey, A., et al. (2009). The impact of cross-border reproductive care or 'fertility tourism' on NHS maternity services. *BJOG, 116*, 1520–1523. doi:10.1111/j.1471-0528.2009.02294.x.

Medhekar, A. (2013). Information search to travel abroad for medical treatment: An empirical study. In *Proceedings of the seventh international conference on health care systems & global business issues*, 3–6 January. Jaipur National University, Jaipur, India.

Mitchener, K. J., & Weidenmier, M. (2008). Trade and empire. *The Economic Journal, 118*, 1805–1834.

Miyagi, K., et al. (2011). The unwritten price of cosmetic tourism: An observational study and cost analysis. *Journal of Plastic, Reconstructive and Aesthetic Surgery, 65*(1), 22–28. doi:10.1016/j.bjps.2011.07.027.

Muscat, N., et al. (2006). Sharing capacities: Malta and the United Kingdom. In M. Rosenmöller, M. McKee, & R. Baeten (Eds.), *Patient mobility in the European Union: Learning from experience*. Copenhagen: European Observatory on Health Systems and Policies.

Newdick, N., & Smith, J. (2010). The structure and organisation of the NHS. Nuffiled Trust.

NHS Choices (2012). Seeking medical treatment in Europe. NHS choices website. http://www.nhs.uk/NHSEngland/Healthcareabroad/plannedtreatment/Pages/Introduction.aspx. Accessed 29 July 2013.

Nuffield Trust (2013). NHS in numbers: Referral to treatment waiting times. Nuffield trust. http://www.nuffieldtrust.org.uk/data-and-charts/waiting-times-hospital-treatment-england. Accessed July 2013.

OECD. (2010). Improving value in health care measuring quality.

OECD. (2010). Policy brief on the international migration of health workforce. OECD/WHO. http://www.oecd.org/migration/mig/44783473.pdf. Accessed 29 July 2013.

OECD. (2009). Quality indicators. http://www.pathqualityproject.eu. Accessed 29 July 2013.

OECD. (2012). OECD health data: How does the United Kingdom compare? http://www.oecd.org/health/healthdata. Accessed 29 July 2013.

Office for National Statistics. (ONS) (2011) *International Passenger Survey, 2011*. Newport: ONS. http://www.ons.gov.uk/ons/dcp171776_274235.pdf. Accessed 29 July 2013.

Pérotina, V., et al. (2013). Does hospital ownership affect patient experience? An investigation into public–private sector differences in England. *Journal of Health Economics, 32*, 633–646.

Peterssen, N. P., & Osterhammel, J. (2005). *Globalisation: A short history*. Princeton: Princeton University Press.

Powell, T. (2012). *NHS charges for overseas visitors standard note*. House of Commons Library. http://www.parliament.uk/briefing-papers/SN03051.pdf. Accessed 29 July 2013.

Propper, C. (2012). Competition, incentives and the English NHS. *Health Economics, 21*, 33–40. doi:10.1002/hec.1804.

Robertson, R., & Burge, P. (2011) The impact of patient choice on equity: Analysis of a patient survey. *Journal of Health Services Research and Policy 16*(S1), 22–28. doi:10.1258/jhsrp.2010.010084.
Siciliani, L., et al. (2013). Differences in length of stay for hip replacement between public hospitals, specialised treatment centres and private providers: Selection or efficiency? *Health Economics, 22*, 234–242.
Smith, R., et al. (2011). Medical tourism: A review of the literature and analysis of a role for bi-lateral trade. *Health Policy, 103*, 276–282. doi:10.1016/j.healthpol.2011.06.009.
Smith, P. C., & Street, A. D. (2013). On the uses of routine patient-reported health outcome data. *Health Economics, 22*(2), 119–131. Accessed 29 July 2013.
Smith, R. (2010). In P. Smith, & S. Glied (Eds.), *Oxford handbook of health economics*. Oxford: Oxford University Press.
Smith, R. D., et al. (2009). Trade and health: An agenda for action. *The Lancet, 373*, 768–773. doi:10.1016/S0140-6736(08)61780-8.
Travis, A. (2013). Immigrants face £1,000 health care levy. The Guardian. http://www.guardian.co.uk/uk-news/2013/jul/03/immigrants-levy-healthcare. Accessed 29 July 2013.
Van Ginneken, E., & Busse, R. (2011). Cross-border health care data. In M. Wismar, W. Palm, J. Figueras, K. Ernst, & E. van Ginneken (Eds.), *Cross-border health care in the European Union*. The European observatory on health systems and policies, Copenhagen.
Wadsworth, J. (2012). Musn't grumble: Immigration, health and health service use in the UK and Germany. IZA. http://ftp.iza.org/dp6838.pdf . Accessed 29 July 2013.
Wanless, D. (2002). *Securing our future health: taking a long term view* (final report). London: HM Treasury. http://webarchive.nationalarchives.gov.uk/, http://www.hm-treasury.gov.uk/consult_wanless_final.htm. Accessed 29 July 2013.
Watkins, P. (2012). National Assembly of Wales. Cross-border health care Quick Guide April 2012. http://www.assemblywales.org/qg12-0012.pdf. Accessed 29 July 2013.
Wennberg, J. E. (2011). Time to tackle unwarranted variations in practice. British Medical Journal, *342,* 687–690. doi: 10.1136/bmj.d1513. Accessed 31 July 2013.
Williams, B., et al. (2000). Private funding of elective hospital treatment in England and Wales, 1997–1998: National survey. *British Medical Journal, 320,* 904–905. doi: 10.1136/bmj.320.7239.904. Accessed 29 July 2013.
Williams, Z. (2013). NHS: Poles, paracetamol and the myth of health tourism. The Guardian, Friday May 17, 2013.
Wilson, E. (2001). Sick of waiting for your NHS op? Then why not go abroad? The Guardian.
Youngman, I. (2009). Medical tourism statistics: Why McKinsey has got it wrong. *International Medical Travel Journal.* http://www.imtjonline.com/articles/2009/mckinsey-wrong-medical-travel/. Accessed 29 July 2013.

What Drives Patient Mobility Across Italian Regions? Evidence from Hospital Discharge Data

Silvia Balia, Rinaldo Brau and Emanuela Marrocu

Abstract This chapter examines patient mobility across Italian regions using data on hospital discharges that occurred in 2008. The econometric analysis is based on Origin–Destination (OD) flow data. Since patient mobility is a crucial phenomenon in contexts of hospital competition based on quality and driven by patient choice, as is the case in Italy, it is crucial to understand its determinants. What makes the Italian case more interesting is the decentralization of the National Health Service that yields large regional variation in patient flows in favor of Centre-Northern regions, which typically are 'net exporters' of hospital treatments. We present results from gravity models estimated using count data estimators, for total and specific types of flows (ordinary admissions, surgical DRGs and medical DRGs). We model cross-section dependence by specifically including features other than geographical distance for OD pairs, such as past migration flows and the share of surgical DRGs. Most of the explanatory variables exhibit the expected effect, with distance and GDP per capita at origin showing a negative impact on patient outflows. Past migrations and indicators of performance at destination are effective determinants of patient mobility. Moreover, we find evidence of regional externalities due to spatial proximity effects at both origin and destination.

Keywords Hospital admissions · Bilateral flows · Gravity model · Cross-section dependence

JEL Classification I11 · C21 · H4 · D12 · R1

S. Balia (✉) · R. Brau · E. Marrocu
Department of Economics and Business, University of Cagliari and CRENoS, Viale S. Ignazio 78, 09123 Cagliari, Italy
e-mail: silvia.balia@unica.it

R. Brau
e-mail: brau@unica.it

E. Marrocu
e-mail: emarrocu@unica.it

R. Levaggi and M. Montefiori (eds.), *Health Care Provision and Patient Mobility*, Developments in Health Economics and Public Policy 12,
DOI: 10.1007/978-88-470-5480-6_6,

1 Introduction

The Italian National Health Service (NHS) is a regionally decentralized tax-funded system where patients are entitled to choose their own preferred provider of hospital and specialized care. The system is characterized by prospective payment schemes and some degree of competition between providers of health care services. Since the constitutional reform approved in 2001, Italian regions have gained a larger autonomy in the administration and organisation of the health care service. The decentralization process has also involved the funding system, which is largely based on regional budgets, and has had effects on the efficiency and effectiveness of the health care provided at the regional level, creating separate and autonomous regional health services (RHSs). Health care provision and financing have become one of the central issues in the debate on fiscal federalism in Italy, because of the consequences that they can have on universalism and equity in health care. In this context, patient mobility concerns both patients' choices and financial resources flows across regions.

Patient decision to be hospitalised in any public or private accredited hospital is mainly determined by health needs and hospital specialization. Patients essentially move to get the best hospital treatments that the NHS can provide, or at least better services than those offered locally. They are expected to move when waiting lists are too long in the local hospitals but also when the perceived quality of their RHS is low. Since hospital in-patient services are provided free at the point of use, patient choice does not depend on treatment costs. At the same time, however, patient mobility is likely to be limited by distance from home, travel costs and the difficulties of getting informal support by the relatives and friends when admission is in an extra-regional hospital.

In this context, the importance of patient mobility is twofold. First, the geography of patient mobility yields very important indications on the actual level of the services provided. When individual preferences are not perfectly observable, "exit" mechanisms-such as patient outflows-reveal the effectiveness of public health care supply. Such information may be adequately exploited by central planners interested in correcting inefficiencies of decentralized provision as well as by local governments wishing to improve the RHSs performance. Second, flow imbalances across regions may challenge the stability of regional budgets. As a matter of fact, public hospitals are financed on the basis of the capitated prospective budget attributed to each Local Health Authority (LHA) and, similarly to private accredited hospitals, they are reimbursed for each admission according to a Diagnosis Related Groups (DRG)–tariff system. National tariffs represent the ceilings to reimbursements for hospital admissions and regional variations are allowed downwards to discourage inappropriate admissions. In the case of interregional patient mobility the national tariff is used for reimbursement between LHAs, with an exception for the Autonomous Provinces (P.A.) of Trento and Bolzano, while intraregional mobility is paid on the basis of regional tariffs. On the one hand, this provides a financial incentive for each RHS to attract patients from outside the region, thus reducing the unit cost associated with each admission (regions that experience high patient inflows rapidly absorb fixed costs). On the other hand, patient mobility also provides a strong financial incentive to restrain outflows of patients who seek treatments delivered also in the region of residence, particularly where the regional tariff is lower than the national one.

Typically, Centre-Northern regions in Italy are net exporters of hospital treatments. This fact translates into additional amounts of financial resources, generated by the compensation of patient flows, in favor of those regions, and exacerbates the North–South gradient in health care in the Italian NHS. Southern regions typically display the highest negative mobility balances, meaning that they pay a higher amount for hospital care services used by their enrolees outside the region than the amount received as reimbursement from the other regions. In 2008, the mobility balance in the South of Italy was between 1.3 (in Abruzzo) and 6.8 (in Calabria) % of regional public health expenditures, thus increasing the deficit of those RHSs. Only one Southern region, Molise, appears to exploit patient flows and reduce its deficit by about 40 %. The RHSs that get the greatest advantage from interregional mobility and transform their deficits into net gains are Lombardia and Emilia-Romagna (where the mobility balances are 2.7 and 4.3 % of regional public health expenditures). These figures reflect the effects of competition due to patient mobility on the sustainability of the RHSs. These stylized facts should also induce the regional governments and LHAs to increase investments aimed at improving the quality and the efficiency of the regional hospital sector. This would decrease the social costs associated with mobility and would also permit to attract more patients, and therefore financial resources, from outside the region. As interregional mobility for hospital care is becoming one important source of reallocation of financial resources between regions, identifying the main determinants of patient flows becomes a central issue.

This chapter investigates patient mobility across Italian RHSs, in a context of regional competition driven by patient choice, using data from hospital discharge records available from the Ministry of Health for 2008. The econometric analysis is based on Origin-to-Destination (OD) flow data. We present results from a gravity model for bilateral patient flows that take into account the count nature of data. We focus on both demand and supply factors in each region and distinguish between pull factors which describe the ability of a region to attract patients from other regions, and push factors which describe the ability to restrain the outflows. We control for the well-known cross-section dependence feature of flow data by including regional pairs' covariates. Besides geographical distance, we consider past migration flows and the share of surgical admissions, which are expected to approximate information and support networks. Externality effects from neighboring RHSs are taken into account by means of spatial lags of some specific pull and push factors.

The chapter is organised as follows. Section 2 presents a selective review of the previous studies most directly related to our contribution. In Sect. 3 we present a detailed description of the Italian regional patient flows along with a brief discussion on the main features of the explanatory variables included in the econometric analysis, which follows in Sect. 4. Section 5 concludes.

2 Related Literature

The empirical literature on patient mobility has been mainly developed in the last fifteen years, with a particular focus on modelling hospital choices and flows across different jurisdictions. A large part of the literature analyses the relationship between competition and quality in health care markets, where patient choice—and therefore patient flows—typically mirrors differences in quality among providers.[1] In an extensive review of the literature, Gaynor and Town (2012) find that the effect of competition on health care quality is generally positive in the case of price-regulated markets. Nonetheless, national health systems often impose severe restrictions on patients' freedom of choice. Where reforms towards patient choice have been put in place, results are consistent with quality improvements, as in the English NHS reform on patient choice accomplished in 2008 (Propper 2012; Bloom et al. 2010; Cooper et al. 2011; Gaynor et al. 2010).

Other works are directly focused on patient mobility aiming at identifying the main drivers of flows occurring within and between health care systems. The related literature presents a clear dichotomy between studies where the unit of analysis is the single patient and macro-level studies where the unit of analysis is a geographical area or a political decision unit. The former have singled out a wide set of potential determinants of mobility (see Victoor et al. 2012, for a survey), referred both to patient characteristics (e.g. education, income and age), and provider characteristics, classifiable as "structure factors" (e.g. availability and accessibility of the providers; cost of treatments), "process factors" (availability of information, continuity of treatment, waiting time and quality of treatment) and "outcome factors" (e.g. mortality indicators).[2] These studies have also detected that a substantial fraction of patients does not consider choice to be very important: they simply choose the nearest provider or follow the advices of their GP, who in turn tends to refer patients to the closest hospitals (Birk and Henriksen 2012).

At the macroeconomic level, the investigation of the determinants of patient mobility has been developed along both the theoretical and empirical level. Theoretical studies often examine patient mobility within a framework of *institutional competition,* where sub-national levels of government (the decision units in a federal system) compete among them. Brekke et al. (2011) consider a situation where regional governments differ in their ability to provide public services

[1] Some studies measure competition using actual and predicted patient flows to build indicators of market structure (Cookson et al. 2013; Gaynor et al. 2010).

[2] Somewhat surprisingly, process factors are found to play a major role, and outcome factors a residual one (Rademakers et al. 2011).

(namely health care). In this context, patient mobility within a system which centrally defines the quality of services, can play a major role in improving overall welfare, though with asymmetric effects: since decentralization creates more incentives for quality improvement, all patients from "high-skill" regions are better off than in a system without mobility. Conversely, in "low-skill" regions only patients who move to a high-skill region benefit from the higher quality provided. Levaggi and Menoncin (2008, 2012) provide another explanation of the large patient flows observed in decentralized health care systems with a central government defining health targets. In the presence of a national regulator characterized by a "soft budget constraint" *vis à vis* the local authorities,[3] inefficient regions have an incentive to send patients to more efficient ones. This increases the probability of being bailed-out by the central authority, thus avoiding the need to finance budget deficits by increasing local taxes. Efficient regions benefit from incoming patients, and compensate the bail-out cost of the service through the soft budget constraint. Hence, part of patient mobility is induced by lack of coordination between competing constituencies in a fiscal federalism framework.

The applied macroeconomic analysis of patient mobility between "competing" constituencies has mainly been based on gravity models, more commonly encountered in the analysis of bilateral international trade and migration flows. With gravity models, flows are viewed as being positively influenced by "mass" accumulation at origin and destination, and negatively influenced by the distance between the two. Origin-specific variables reflect potential demand or repulsion effects, while destination-specific indicators should proxy the degree of attraction.[4] In the health economics literature, gravity models were originally introduced by Morrill and Earickson (1968) for modelling hospital choice, then subsequently adopted for the analysis of aggregate bilateral flows such as those studied in this chapter. The studies related to our research work estimate models for patient mobility across different regions in Italy and Spain (respectively, Levaggi and Zanola 2004; Cantarero 2006). Levaggi and Zanola investigate the determinants of net patient flows from origin regions to the rest of Italy over the period 1995–1997 and find that regions providing better or faster services are characterized by lower outflows. Cantarero, using data on net patient flows across Spanish regions over the period 1996–1999, finds that patients from the worse-off regions move more than those from regions that provide better health services (quality-driven mobility). Fabbri and Robone (2010) focus on bilateral patient flows for hospital care between LHAs in Italy. They estimate a gravity model for count data and investigate the role of scale economies and the impact of the North–South economic divide on mobility. They control for the effect of push and pull factors

[3] In fact, health care is a public and merit good which the central regulators want to provide equitably in the whole federation.

[4] As remarked by Lippi Bruni et al. (2008), the importance of distance differs significantly among treatments, as demand for health services is far from homogeneous. In particular, patients generally show a greater willingness to travel for major treatments. Hence, the intensity of the distance decay effect cannot be generalized.

related to both origin and destination LHAs, geographical factors, such as measures of actual distance and contiguity between LHAs, and spatial factors based on weighted distances between LHAs. Their main findings suggest that richer LHAs tend to attract more patients, especially the most severe cases, and LHAs in the South are less able to "import" patients and restrain outflows.

From a methodological viewpoint, the health economic literature has been only recently enriched by approaches based on the estimation of gravity specifications augmented by spatial features. In this chapter, we thus aim to contribute to the current debate on the determinants of patient flows by focusing on the regional dimension of the phenomena, given its crucial implications on regional budgets and, ultimately, on patients utility.

3 Data

3.1 Patient Mobility in Italy

We use data on hospital discharges that occurred in 2008 in all public and private accredited hospitals of the 21 RHSs that provide health care in the country.[5] Information on in-patient care is collected by each hospital at the time of discharge and regularly transmitted through the LHA to the Ministry of Health. Each admission episode is classified using version 19 of the US DRG system, and length of stay is reported. The data at hand contain information that allow us to identify the hospital of admission, the corresponding LHA and region, as well as the LHA and the region of residence associated with each single discharge. The total number of hospitals with complete discharge data is 1,304. This corresponds to a total number of 11,701,920 admission episodes, of which 29.5 % are identified as flows because they either occur in a different region (inter-regional mobility) or in the region of residence but not in the LHA where the patient is registered (intra-regional mobility).

In this chapter we focus on inter-regional patient mobility because this represents an important source of external revenue for regions that are able to attract patients from other regions (net exporters). For the 21 Italian regions, we construct a square matrix that describes inter-regional patient flows by aggregating the number of admissions of patients from each possible region of origin (enrolees in RHS *i*) in public or private accredited hospitals of each possible region of destination (RHS *j*). For the purpose of our analysis, the main diagonal of the matrix is

[5] Approximately 33.6 % of public hospitals are run by LHAs; 11 % are autonomous public enterprises; 4.4 % are scientific institutes for research, hospitalization and health care and 2 % are medical school hospitals. Private accredited hospitals represent about 45 % of the total number of providers but only 16 % of total admissions.

set to 0 in order to exclude intra-regional flows. This leaves us with 420 bilateral OD patient flows.

We concentrate our attention on potentially avoidable mobility, thus we exclude admissions that are classified in the four Major Diagnostic Categories (MDC) which most likely represent unavoidable mobility. These MDCs refer to traumatology, HIV and burns. The total number of admission episodes outside the region of origin is 863,953, which is 25.1 % of total mobility. About 47.2 % of these episodes have a surgical DRG (and hence, on average, are compensated on the basis of higher national tariffs than admissions with medical DRGs) and 64.4 % are classified as ordinary admission.[6] About 44.3 % of total OD flows are generated by Southern regions (382,682 admission episodes). About 17.2 % of total flows are from regions in the South to regions in the North (148,550 admission episodes), while only 2.8 % of flows are from the North to the South (23,831). Flows between Northern origins and destinations represent about 29.5 % of total inter-regional mobility, while this figure is only 10.9 % in the case of flows between Southern origins and destinations.

Table 1 illustrates the main differences between Italian regions in terms of their contribution to inter-regional mobility. We calculate the creation rate as the percentage ratio between enrolees of region i admitted in hospitals of other regions and the total number of patient flows in Italy, and the attraction rate as the percentage ratio between non-enrolees admitted in hospitals of region j and the total number of patient flows in Italy. Both rates show different spatial patterns. The regions that create more inter-regional mobility are the most densely populated territories in the country: Campania, in the South, generates about 10.6 % of total flows, followed by Lombardia, in the North (8.7 %), and Lazio, in the center of Italy (8.4 %). The least populated regions (P.A. Trento, Friuli Venezia-Giulia, Molise, Valle d'Aosta and P.A Bolzano) and Sardegna (probably due to insularity) generate less than 2 % of total flows. The regions that admit more non-resident patients are Lombardia, Emilia-Romagna and Lazio with attraction rates of about 19, 13.5 and 13 %, respectively. The attraction rate is largely higher in the Central-Northern regions, 81.8, than in the Southern regions, 18.2.

We further examine patient flows in each region using the mobility index. This index measures the ratio between the inflow rate and the outflow rate, and takes values larger than 1 when the RHS is a net importer of patients from other RHS, thus being able to offset the outflows with larger inflows.[7] The maps reported in Fig. 1 confirm the role of spatial proximity and show a clear North–South gradient. For most of the regions in the Centre-North of Italy the mobility index is larger

[6] Ordinary admissions imply at least one night spent at the hospital and exclude admissions in long-term care and rehabilitation wards, as well as admissions of healthy babies born at the hospital.

[7] The inflow rate is the percentage ratio between non-enrolees admitted in region j (inflows) and the total number of admissions in region j. The outflow rate is the percentage ratio between enrolees of region i admitted in other regions (outflows) and the total number of admissions of enrolees of region i.

Table 1 Indicators of inter-regional patient mobility

Regions	Creation rate	Attraction rate	Outflow rate	Inflow rate	Mobility index
Piemonte	6.54	5.67	6.97	6.10	0.87
Valle d'Aosta	0.63	0.24	21.78	9.44	0.43
Lombardia	8.71	18.71	4.20	8.61	2.05
P.A. Bolzano	0.51	0.87	4.27	7.02	1.64
P.A. Trento	1.72	1.20	15.72	11.47	0.73
Veneto	5.73	7.52	6.20	7.98	1.29
Friuli Venezia-Giulia	1.67	2.36	7.10	9.77	1.38
Liguria	4.91	4.66	11.51	10.98	0.95
Emilia-Romagna	5.91	13.54	6.43	13.59	2.11
Toscana	4.75	8.11	6.60	10.75	1.63
Umbria	2.44	2.98	12.63	14.98	1.19
Marche	3.78	3.35	11.08	9.93	0.90
Lazio	8.39	12.61	6.13	8.94	1.46
Abruzzo	4.72	3.87	14.55	12.24	0.84
Molise	1.65	2.37	19.26	25.51	1.32
Campania	10.61	3.07	6.97	2.12	0.30
Puglia	7.75	3.25	7.56	3.32	0.44
Basilicata	2.94	1.75	21.49	13.99	0.65
Calabria	7.86	1.18	15.61	2.70	0.17
Sicilia	6.84	2.20	5.80	1.94	0.33
Sardegna	1.93	0.53	5.20	1.49	0.29
South	44.29	18.21	8.61	3.73	0.43
Centre-North	55.71	81.79	6.63	9.45	1.42

than 1 (larger than 2 for Emilia-Romagna and Lombardia), while all Southern regions display indexes lower than 0.7. Spatial patterns in hospital admissions, which are likely due to the influence of demand and supply features of the RHSs at origin and destination, seem to reflect the well-known North–South economic divide, as richer and better equipped regions tend to attract more patients and resources.

3.2 Origin and Destination Characteristics

Inter-regional patient flows depend on characteristics of both origin and destination regions, which are related with economic and demographic aspects and features of the hospital care services. We have selected the set of variables which are expected to influence patients' choice about the hospital where to seek care, as well as the ability of the RHS to attract inflows and restrain outflows of patients. The complete description of all variables is reported in Table 2. In this section, we

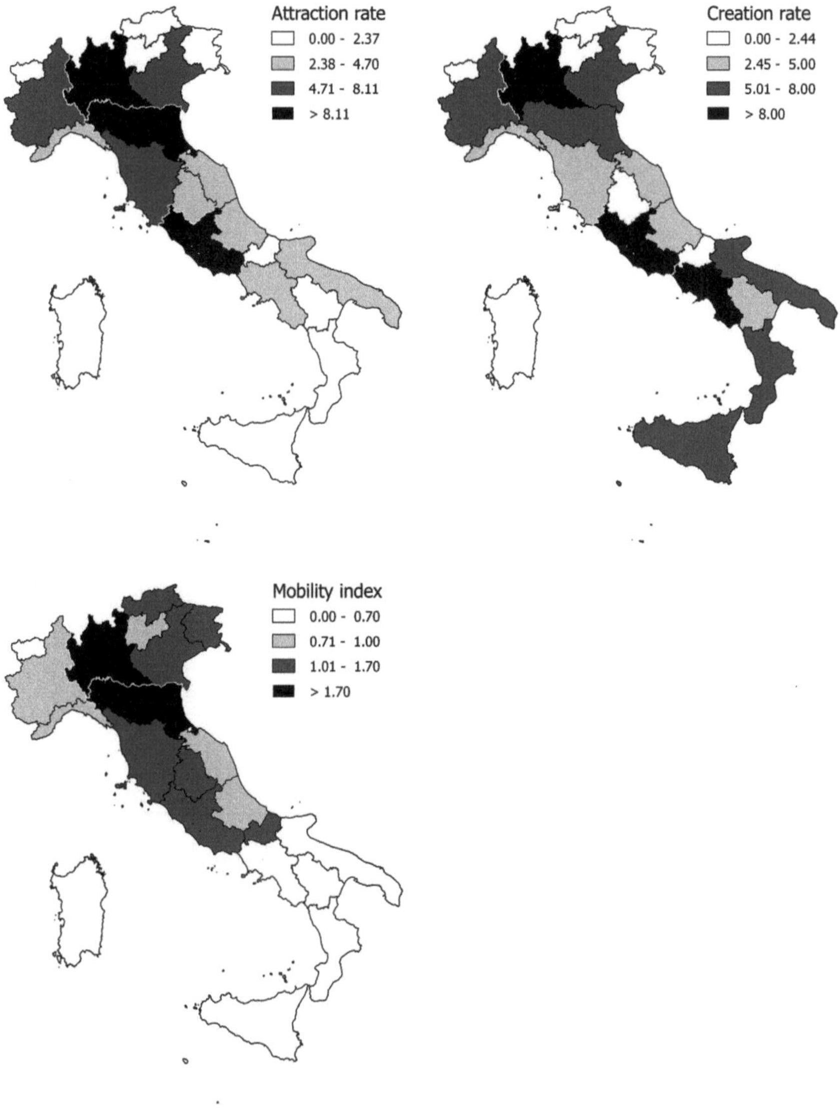

Fig. 1 Spatial pattern of inter-regional patient mobility

distinguish between push and pull factors, which play a different role at origin and destination, respectively.

Origin Characteristics

In the following we motivate the choice of push factors used in the empirical model.

Table 2 Descriptive statistics, variable definitions and data sources

Variable	Mean	SD	Min	Max	Definition	Primary source	Year
Patient flows	2,057.03	4,093.96	0	32,675	Hospital admissions of patients from origin region *i* in destination region *j*	Italian ministry of health	2008
GDP pc	20,855	5,105	13,507	28,116	Per capita GDP (millions of euros), constant values (2000)	ISTAT	2008
Population	2,859,289	2,487,688	127,065	9,742,676	Resident population (annual average)	ISTAT	2008
Population aged 0–14 (%)	13.78	1.44	11.34	16.73	Share of the population aged 0–14 years old	ISTAT	2008
Population aged over 65 (%)	20.64	2.52	15.87	26.78	Share of the population aged 65 years old or over	ISTAT	2008
Hospital capacity	4.27	0.51	3.62	5.80	Number of hospital beds per 1,000 inhabitants	NHS statistical yearbook	2008
Organisational efficiency	0.40	0.41	0.01	1.80	Distance between hospital capacity and the national target (absolute value)	Own calculations on NHS statistical yearbook	2008
HECC dummy	0.14	0.36	0	1	High efficiency in complex cases: 1 if the region has high complex cases and high performance in lenght of stays, ordinary admissions; 0 for all the other regions	Italian ministry of health	2008
Technology index	0.004	0.98	−1.34	2.70	Factor scores for high technology medical equipment per 1,000,000 inhabitants	NHS statistical yearbook	2008
Surgical admissions (%)	2.33	6.79	0	73.48	Share of surgical DRG admissions from origin *j* over surgical DRG admissions in destination *i*	Italian ministry of health	2008
Migration flows	3,993	6,412	0	44,756	Residential changes of citizens from origin *i* to destination *j*	Eurostat	2000–2004
Accessibility	2.95	0.80	2	5	Potential accessibility by road, train, air and time to the market; five groups (from 1 = very low, to 5 = very high)	Espon project 2.4.2	2006
Distance	469.01	248.35	54.50	1,125.50	Distance in Km between the centroids	Own calculations	

Per capita GDP. Per capita income may pick up two different effects: at the micro level, its effect can be interpreted as patients' ability to travel and seek care outside the region of residence (richer individuals choose better hospitals regardless of travel and accommodation costs); at the macro level, the income effect can be interpreted as a greater ability of the regional health care system to provide hospital services (poorer regions experience outflows of patients towards richer regions).

Population. Population indicates the number of enrolees to the RHS and approximates the internal demand for health care. Bigger regions have a higher internal demand of hospital care, this might induce more variety in the range of specialized health services provided in the area. Furthermore, higher populated regions may exploit economies of scale leading to cost minimization, more and better services. For this reason, highly populated regions should be able to restrain patients outflows better than small regions.

Population age 0–14 and *Population over 65*. This two demographic indicators allow us to capture the effect of aging on the likelihood to seek care in extra-region hospitals. Since morbidity rates among the youngest individuals are typically lower than at adult ages, we expect a negative impact on outflows. Likewise, we also expect a negative impact for the oldest individuals in the population since they are usually less prone to travel.

Technology index. The technological endowment of each RHS is captured by information on the number of high technology medical equipment. We consider ten different devices and compute a synthetic technological index on the basis of principal factor analysis, scoring the first factor, that accounts for approximately 53 % of the variance, by the regression method.[8] We expect that the higher the index, the smaller the patients' outflows.

Hospital capacity. The number of hospital beds per 1,000 inhabitants is an indicator of hospitals supply in the region. The lower the availability of beds, the higher the waiting times for admissions and the larger the outflows. The downside of this indicator is that it does not capture the quality of care.

Organisational efficiency. An excess of beds in hospitals, however, is typically considered a signal of bad management, and can translate into a waste of resources. We propose a measure of the distance (in absolute value) between the actual number of hospital beds per 1,000 inhabitants and the national target on the optimal number of beds (4 per 1,000 inhabitants), recommended by the Italian Ministry of Health guidelines for the period 2007–2009. Although this indicator represents a single dimension of hospital organisational efficiency, we can assume that it is a fairly good approximation of organisational efficiency in the regions. Inefficient RHS are further away from the target (they can be either over- or under-

[8] The devices are magnetic resonance imaging, linear accelerator in radiotherapy, computed axial tomography (CT), gamma CT, mammogram, gamma camera, positron emission tomography (PET) or CT-integrated PET, hyperbaric chamber, digital angiography, automated immunochemistry analyzer.

endowed with respect to the target) and are expected to experience larger patients' outflows.

HECC dummy. The "high efficiency in complex cases" dummy combines the information related to the case-mix index (CMI) and the performance index (PI), which can be used together to measure hospital efficiency at the regional level. The regional CMI is calculated as the ratio between the average weight of admissions in a region and the average weight of admissions at the national level. The DRG relative weights, which measure the amount of financial and physical resources allocated to DRGs, are used to calculate the average weights. Hence, the CMI indicates the degree of clinical complexity of hospital admissions in each region with respect to the national average and represents an index for specialization. A CMI higher (or equal) to 1 indicates a greater (or equal) clinical complexity compared to the national benchmark. The PI compares the regional performance in managing hospital length of stays and is calculated as the ratio between the case-mix standardized average length of stays in each region and the benchmark level (typically the national average).[9] A PI lower than (or equal to) 1 indicates that, assuming equal complexity, hospital stays are shorter (or have the same length) than at the national level, thus suggesting higher (or at least equal) efficiency relative to the standard. We calculate the CMI and PI on ordinary admissions only (excluding 1-day hospital stays) and we build a comprehensive variable that identifies the most efficient regions as those characterised by high complexity and high performance.[10] High efficiency regions are likely to restrain patients outflows better than other regions.

Destination Characteristics

In the following we motivate the choice of pull factors used in the empirical model.

Per capita GDP. As a pull factor, per capita GDP is expected to have a positive effect on inflows. Richer regions are more attractive because they are expected to provide more efficient and effective health care services.

Population. Bigger regions should be able to attract patients from smaller regions because of the larger variety of specialized care provided. Therefore, the size of the population is expected to be positively related to patients inflows.

Technology index. We expect that the higher the index, the higher the patients' inflows.

Hospitals capacity. Regions with a higher number of beds, and perhaps shorter waiting lists for admissions, are more attractive.

[9] The case-mix standardization controls for regional differences in the complexity of the admissions. Consequently, the indicator of regional standardized average lengths of stays assumes that each region share the same complexity of cases as the national average.

[10] In the yearly report on hospitals activity based on the discharges database, the Italian Ministry of Health assesses the efficiency of RHS on the basis of data on ordinary admissions with stays longer than one day. Stays are typically longer and cases can be more clinically complex in long-term and rehabilitation wards and neonatal care units.

Organisational efficiency. Less efficient regions (under- or over-endowed) should attract less patients from other regions.

HECC dummy. As explained above, efficiency calculated in terms of both clinical complexity of admissions and performance in managing hospital stays is expected to play an important role as a pull factor. High efficiency regions should attract more patients compared to other regions.

Accessibility. Attractiveness might benefit from the patients' ability to access the destination region in an easy way. We use a five-category discrete variable that describes potential accessibility by road, train, air and time to the market. We expect a positive effect on inflows because a higher accessibility should increase the probability that individuals move to seek care in extra-regional hospitals.

Origin–Destination Flow Characteristics

Geographical distance. We include the distance for each pair of destination and origin regions as a proxy of transport costs and thus it is excepted to have a negative impact on patient movements. Within the gravity model framework the geographical distance is also expected to account for spatial correlation among the observational units.

Migration flows. Past inter-regional migration between any two OD regions is expected to have a positive impact on patient mobility. Past migration flows are a powerful source of knowledge for patients seeking health care. People who have migrated may provide their relatives and friends looking for good health care with valuable information on the medical services present in their region of residence and in the contiguous ones. Moreover, patients are more likely to seek hospital care in regions where relatives or friends of their same origin region migrated. The latter can provide informal support before, during and after the hospital stay, especially when treatments require long stays. For this reason, past migration flows allow us to account for spatial correlation in patient mobility.

Surgical Admissions. As another control for the flows between any two OD regions, we include an indicator of the share of patients resident in origin i and admitted with a surgical DRG in region j over the total amount of patients admitted with surgical DRG in the destination region. The national tariffs associated with surgical DRGs are typically higher compared to medical DRGs, due to complexity and effective costs. As a driver of inter-regional patient mobility, we expect a positive effect due to micro level learning (at destination) and communication (at origin) of patients sharing their experiences, within family and friends networks, on treatments received in specific destinations.

4 Empirical Analysis

4.1 Economic Framework

The empirical analysis of the determinants of inter-regional patient flows is based on the estimation of gravity models. The general model is formalized as follows:

$$y_{ij} = \alpha + X_{oi}\beta_o + X_{dj}\beta_d + X_{ij}\beta_{od} + u_{ij} \tag{1}$$

where y_{ij} is the number of patients resident in region i that chose to be treated in hospitals located in region j in the year 2008. The total number of possible inter-regional flows is 420. The matrix X_{oi} comprises the variables which are supposed to represent the origin's features, while the matrix X_{dj} includes variables for the destinations' characteristics. These two sets of regressors are thus expected to account for the emissiveness and attractiveness traits of each region, as detailed in the previous section. The matrix X_{ij} includes the variables which pertain to the pair of regions involved in each patients flow, namely the geographical distance, past migration flows and the share of surgical admissions; conditional on correct model specification, the u_{ij} term is expected to be an i.i.d. process.

Since our response variable is a count variable the natural starting point is to consider the Poisson regression model, which entails specifying the conditional mean (μ_{ij}) as a function of a set of covariates X. The standard parameterization of the conditional mean is $E(y_{ij}|X) = \mu_{ij} = exp(X\beta)$ to ensure that the non-negativity constraints are not violated. Since the Poisson distribution is characterized by the equidispersion property, the variance is equal to the mean so that the Poisson model is intrinsically heterosckedastic. The equidispersion property has often been found to be excessively restrictive in empirical applications because count data are usually characterized by overdispersion. Our sample data is no exception, our dependent variable exhibits a mean value of 2,057.03 and a standard deviation of 4,093.96.

Unobserved heterogeneity is deemed to be one of the most common causes of overdispersion (Cameron and Trivedi 2005), which can also yield an excessive number of zero observations.[11] It is common to deal with such heterogeneity by considering a continuous mixture of the Poisson distribution. This requires modifying the specification of the mean as $E(y_{ij}|X) = \mu_{ij}\eta_{ij}$, with μ_{ij} defined as before and η_{ij} a random term with $E(\eta_{ij}) = 1$. In this case, the Poisson mixture has the same mean as the original Poisson. When η_{ij} follows the gamma distribution with variance α, the negative binomial model results. The first two moments are $E(y_{ij}|X) = \mu_{ij}$ and $Var(y_{ij}|X) = \mu_{ij} + \alpha\mu_{ij}^2$, and α is the over-dispersion parameter to be estimated.[12] Since our flow data are characterized by over-dispersion, the

[11] In our sample data we just have one pair of regions, Val d'Aosta and Basilicata, featuring a zero patient flow.

preferred functional form chosen to estimate the gravity relation reported in Eq. (1) is the negative binomial model.[13]

When analysing flow data, another relevant issue is the existence of spatial dependence (Griffith and Jones 1980; Le Sage and Pace 2008, 2009). Flows from a given origin are usually related to the degree of emissiveness of neighboring regions; likewise, flows towards a specific destination are enhanced or reduced according to the degree of attractiveness of nearby destination locations. In this study we deal with the spatial dependence issue by including regressors characterized by spatial features (geographical distance, regional accessibility, spatial lags of some explanatory variables) and regressors which allow us to properly account for the characteristics of the OD regional pairs. We thus include the share of surgical admissions and past migration flows.[14]

4.2 Results

The main results of the analysis of the Italian inter-regional patient flows determinants are reported in Table 3. In the first column we present our baseline model, which includes—for both origin and destination regions—population, per capita GDP and hospital capacity; origin features are also accounted for by including the share of population in the groups aged 0–15 or above 65 years. In the base specification, the regional pairs' characteristics are captured by the geographical distance. The estimated coefficients are in line with the effects expected on the basis of the arguments presented in Sect. 3. Population, acting as the "mass" variable usually included in gravity specifications, turns out to be more effective as a pull rather than a push factor (the estimated elasticity are 1.01 and 0.89 respectively). Conversely, GDP per capita is effective in restraining patient outflows; in absolute terms the estimated elasticity at origin is more than twice the one at destination (-1.88 vs. 0.79). Hence, we find evidence that the macro level effect proxied by GDP per capita described in Sect. 3.2 dominates the micro level effect. Supply factors, represented by hospital capacity, play a significant role only at destination. The additional demand side controls exhibit the expected negative sign on patients mobility because population in the "dependency" age classes are

[12] Note that this specification is referred to as negative binomial 2 (NegBin2); the negative binomial 1 entails a linear variance function. The NegBin2 specification is typically preferred because the quadratic form has been proven to provide a very good approximation to more general variance functions. This is a remarkable advantage because the maximum likelihood estimators for negative binomial models are not consistent when the variance specification is incorrect.

[13] In a preliminary analysis we also estimated the Poisson model, but we found overwhelming evidence in contrast with the equidispersion assumption.

[14] Fabbri and Robone (2010) claim that past residential migration is one of the most relevant variable that can generate network autocorrelation in patient flows.

Table 3 Determinants of inter-regional patient flows

Negative binomial models	1	2	3	4	5	6	7	8	9
Origin									
GDP pc	−1.876***	−1.788***	−1.771***	−1.774***	−1.723***	−1.519***	−1.454***	−1.581***	−1.536***
Population	0.889***	0.775***	0.607***	0.612***	0.636***	0.688***	0.703***	0.599***	0.740***
Population aged 0–14 (%)	−0.121	−0.136*	−0.170***	−0.169***	−0.162**	−0.123**	−0.114**	−0.138**	−0.126**
Population aged over 65 (%)	−0.092**	−0.064	−0.072*	−0.071*	−0.067*	−0.069**	−0.058*	−0.088**	−0.063*
Hospital capacity	−0.063	−0.021	−0.011	−0.012	−0.069	−0.361**	−0.320*	−0.227	−0.483***
Organisational efficiency					0.071	0.159	0.175	−0.040	0.345
HECC dummy					−0.085	−0.056	−0.075	0.017	−0.111
Spatial lag–Organisational efficiency						−3.275***	−3.228***	−3.376***	−3.253***
Spatial lag–HECC dummy						−0.458	-0.110	−0.618	−0.240
Destination									
GDP pc	0.787***	0.796***	0.547***	0.238	−0.022	0.722*	0.328	0.597	0.993***
Population	1.096***	1.185***	0.955***	0.908***	0.811***	0.851***	0.782***	0.824***	0.838***
Technology index					0.121*	0.251***	0.251***	0.175**	0.325***
Hospital capacity	0.235***	0.230***	0.193***	0.215***	0.791***	1.452***	1.385***	0.880***	1.949***
Organisational efficiency					−0.780**	−1.795***	−1.634***	−1.056**	−2.426***
HECC dummy					0.494***	0.403***	0.441***	0.721***	0.165
Accessibility				0.148*	0.245***	0.223***	0.307***	0.372***	0.114
Spatial lag–technology index						1.527***	0.953*	0.560	2.256***
Spatial lag–organisational efficiency						5.430***	3.433*	0.860	8.830***
Spatial lag–HECC dummy						−8.873***	−5.808**	−2.386	−13.460***
Origin-destination									
Distance	−1.569***	−1.086***	−1.164***	−1.154***	−1.160***	−1.274***	−1.223***	−1.366***	−1.220***
Surgical admissions (%)		0.097***	0.076***	0.077***	0.073***	0.071***	0.063***	0.080***	0.062***
Migration flows			0.069***	0.068***	0.066***	0.058***	0.057***	0.064***	0.055***
α parameter	0.748***	0.598***	0.529***	0.526***	0.502***	0.451***	0.433***	0.455***	0.464***
Log-likelihood	−3072.0	−3015.9	−2987.3	−2985.8	−2975.2	−2949.7	−2785.3	−2555.8	−2717.2
Pseudo R^2	0.104	0.120	0.129	0.129	0.132	0.140	0.146	0.164	0.142

Number of regional units: 21; total number of observations: 420

Model 1–6 refer to total mobility; Model 7–9 refer, respectively, to mobility for ordinary admissions, surgical DRGs, and medical DRGs

All regressions include a constant; the variables Population, GDP pc and Distance are log-transformed. The spatially lagged variables are obtained by pre-multiplying the relevant variable by the row-standardized inverse distance matrix. The Migration flows coefficient is multiplied by 1,000

Level of significance: *** 1 %, ** 5 %, * 10 %

less inclined to travel. However, only the share of oldest people is statistical significant. Finally, geographical distance is highly significant and exhibits a sizeable discouraging effect (elasticity estimate of −1.57) on patient flows. It is worth noting that the estimate of the variance function parameter ($\alpha = 0.75$) is highly significant, indicating that overdispersion is a relevant feature of the data. This finding is robust with respect to all the other specifications reported in Table 3.

In order to improve the understanding of the patient flows drivers we proceed by subsequently augmenting the baseline model. We first include additional variables for the regional OD pairs and then for the origin and destination regions. This estimation strategy is selected because a better characterization of the OD pairs is expected to clear any remaining cross-section dependence, not fully accounted for by the inclusion of the geographical distance. Therefore, in the regression models (2) and (3) of Table 3, we subsequently include the share of surgical admissions and past migration flows. As stated in Sect. 3, we believe that the share of surgical admission is a good proxy of hospital performance on highly complex treatments received in the destination region. Since it is measured as the share of people from a region who selected a specific destination for these treatments, it is expected to explain the proportion of patient mobility driven by the information diffused at origin by patients who were treated at specific destinations. Likewise, patient mobility should be positively affected by past migration flows given that they represent both a powerful source of knowledge and a reliable support network for arrangements, especially when treatments require long stays for patients seeking health care outside their own region.

According to the models reported in columns (2) and (3) of Table 3, the share of surgical admissions and past migration flows are statistically relevant in accounting for the variability of patient flows. Moreover, the previously discussed results for the baseline model are largely confirmed, although, as expected, the estimated coefficients from models (2) and (3) imply smaller effects. It is worth noting that the latter models yield higher values for the log-likelihood and that the share of the youngest population turns out to be significant in both models.

Model (4) augments model (3) by including the accessibility index at destination. The estimated coefficient (0.148) points out that easy-to-access locations are significantly more attractive to patients seeking hospital care in other regions, even when transport cost have been already accounted for by geographical distance. Note that in model (4) per capita GDP at destination is no longer significant. This is likely to be due to collinearity issues because, in general, we expect territories with a high degree of accessibility (implying for instance the existence of widespread and advanced transport infrastructures) to be located in the wealthiest regions.[15]

[15] In order to account for the particular morphology of the Italian territory, an undisputed factor in driving the well-known North–South divide, we also include a dummy variable to discriminate between Centre-Northern and Southern regions. The dummy turned out to be highly significant with the expected negative sign, but it makes negative the coefficient of per capita GDP. This is

In models (5) we include the two efficiency indicators at both origin and destination and the technology index at destination. The latter is positive (estimated coefficient 0.12) and significant at conventional levels, confirming that RHS endowed with advanced technological instruments are able to attract more patients inflows. A unit change in the overall technology index implies an increase of nearly 250 non-resident patients. This finding supports the idea that attractiveness can increase due to higher investments in advanced medical devices.

The efficiency indicators do not prove to be relevant factors in restraining patient outflows as they both turn out to be not significant at origin. However, they are highly significant at destination. This indicates that regions which have an RHS characterized by organizational inefficiency (coefficient estimate −0.78) are less attractive for potential patients, and vice versa. Organizational inefficiency at the regional level may be due both to single non-adequately-sized hospital or to a large number of small hospitals widespread in the regional territory. In general, this might generate high management costs and, due to budget constraints, crowd out resources that could be employed to enhance both quality and variety of the health treatments and technological equipment. Conversely, we found the opposite effect (coefficient estimate 0.50) for regions which are characterized by high efficiency in terms of both the case-mix complexity and the performance in hospital stays. Note that the inclusion of the additional technological and efficiency indicators leave largely unchanged the estimated effects and significance of the previously included explanatory variables.

In the final stage of our analysis we address the issue of possible externality effects produced by neighboring regions at both origin and destination.[16] We test for the existence of such effects by including the spatial lag of some specific regressors, namely the technology index and the organizational and efficiency indexes.[17] Each spatial lag is obtained as the weighted average of the corresponding regional regressors values, with weights proportional to the inverse of the geographical distance between any two regions. Focusing on the results of model (6), we find that, at destination, the spatially lagged terms of the technology index and of the two efficiency indicators are highly significant and with the expected sign, thus indicating that being located close to efficient and technological advanced RHS enhances the attractiveness of a given region. This result is likely to be driven by two different, but possibly complementary, mechanisms. The first one is related to the learning and communication processes described above: individuals who for some reason (migration, journeys, previous health treatments)

(Footnote 15 continued)
due to the fact that the two variables are significantly negatively correlated. For these reasons we prefer the specification that includes per capita GDP only.

[16] The analysis of the complete pattern of spatial interactions along the lines suggested in Le Sage and Pace (2008, 2009) goes beyond the scope of this study and thus is left for future research.

[17] The selection of the lagged regressors was carried out on the basis of a preliminary analysis performed on all the origin and destination covariates.

happen to acquire knowledge on the reputation of hospitals in a certain area spread this information in their origin region, thus increasing the flows from that region to the destination surrounded by efficient RHSs. The other explanation may be related to a competitive market kind of mechanism: a region located close to health services efficient areas has an incentive to emulate its neighbors in providing high quality treatments, contending extra-regional patients to nearby regions.

The latter mechanism may also explain the results found for the spatially lagged terms at the origin. The more efficient the neighbors of given region, the higher the incentives for that specific region to increase its own level of efficiency in order to restrain outflows. This is reflected in the sign of the coefficients for the high efficiency dummy at the origin, although only the organizational efficiency coefficient turns out to be significant. In this case, lower efficiency of neighboring regions would not induce any special effort at origin in improving efficiency, however it would reduce the average outflows because of a reduction in flows towards closer regions. It is relevant to remark that the non-spatially lagged regressors proved robust to the inclusion of the additional geographical covariates both in terms of magnitude and statistical significance.[18]

In order to test the robustness of our results, we re-estimate model (6), which is the most general specification, for a subset of specific flows, namely those related to ordinary admissions, surgical DRGs and medical DRGs. The results are reported in columns 7–9 of Table 3. Although, as expected, the magnitude of the estimated coefficient varies across the estimated models, the results previously discussed with respect to model (6) are generally confirmed. The only noticeable exception is represented by the spatial lags at destination for surgical DRGs flows, which are no longer significant. This result is in line with the intrinsic characteristics of treatment complexity, hospital specialization patterns and patient mobility motivations. For highly complex cases, patients tend to select the hospitals specialized in those cases, because they expect these to provide the highest quality treatment and thus to enhance the probability of a successful outcome. In this selection process, only very specific knowledge is applied and this is even more so when extra-regional destinations are considered. This is the reason why information on RHSs close to the selected destination becomes much less valuable yielding non-significant effects.

Overall, the results presented in this analysis provide interesting insights into the determinants of patients flows in the Italian NHS, characterized by an enhanced citizens opportunity to seek health treatments all over the national territory and framed within a persistent and growing North–South divide, that translates into differently performing RHSs. This becomes even more relevant in a context of centrally-defined targets and low variability in per capita public health expenditures across regions.

---

[18] Moreover, note that the "hospital capacity" variable turn out to be significant in model (6).

5 Conclusions

In this chapter we investigate the determinants of inter-regional patient mobility in Italy. This is a relevant issue because the Italian NHS is a regionally decentralized tax-founded system, characterized by patient choice and by a high degree of autonomy for RHSs in designing health care provision at the local level. Understanding which are the most influential drivers of patient mobility is of key importance, as mobility is increasingly determining surplus/deficit of regional budgets and thus the sustainability of the RHSs, with non-negligible consequences in terms of both universalism and equity of health care provision.

We analyze count data from hospital discharge records for inter-regional patient flows in 2008, by estimating negative binomial specifications within a gravity model framework. Our findings indicate that the most effective pull factors are hospital capacity, advanced technological endowment, high degree of efficiency and accessibility of destinations RHSs. Once we control for the demographic structure, we find that at origin patient flows are held back by high levels of per capita GDP, supposed to capture the availability of adequate local health service, and by hospital capacity. The well-known cross-section dependence feature of flow data is accounted for by including variables that characterize the unique traits of the OD pair involved in each flow, namely geographical distance (found to discourage mobility), the share of surgical admissions and past migration flows (with enhancing effects due to information and support networks), and by including spatial lags of the technology, organizational and efficiency indices. The latter detect the existence of positive externalities from proximate regions through learning and communication processes. Information on hospitals' reputation acquired in a certain area, is likely reported at origin, thus increasing the flows from that given origin to the destination location surrounded by efficient RHSs. Evidence of positive externalities is also consistent with competitive market mechanisms because regions emulate their efficient neighbors, thus attracting extra-regional patients. Our main results are confirmed when the analysis is conducted for the sub-sample of ordinary admissions and the sub-groups of DRG (surgical and medical).

Overall, our findings provide sound evidence on what drives inter-regional patient mobility across Italian regions. Our results also call for more attention on the financial sustainability of a decentralized system with centrally defined targets in a context characterized by persistent and growing socio-economic gaps. From a policy perspective, our findings on the significant presence of externalities could suggest that a greater coordination at the central level (for example by means of "fine-tuned" reimbursement schemes) is desirable in order to design or reinforce existing patterns of regional health care specialization so that inter-regional mobility is expected to become mainly specialization-driven rather than induced by the current efficiency gap. In future research, it is worth investigating the potential welfare loss associated with excessive mobility. This is still an open question since, at least in the short run, the inefficient RHSs might benefit from "importing" hospital care from other regions. Therefore, more focused policy prescriptions cannot be directly derived from our analysis.

Acknowledgments The research leading to these results received funding from the Sardinian Region (LR7). The authors would also like to thank Daniela Moro for valuable assistance in preparing the database. We thank Lucia Lispi and Pietro Granella from the Italian Ministry of Health for offering access to the hospital discharge data.

References

Birk, H. O., & Henriksen, L. O. (2012). Which factors decided general practitioners' choice of hospital on behalf of their patients in an area with free choice of public hospital? A questionnaire study. *BMC Health Services Research, 12*(126), 1–10.

Bloom, N., Propper, C., Seiler, S., & Van Reenen, J. (2010). The impact of competition on management quality: evidence from public hospitals. *London School of Economics CEP Discussion Paper 983.*

Brekke, K. R., Levaggi, R., Siciliani, L., & Straume, O. R. (2011). Patient mobility, health care quality and welfare. Norwegian School of Economics and Business Administration Discussion Paper 13/11.

Cameron, A. C., & Trivedi, P. K. (2005). *Microeconometrics: Methods and applications.* New York: Cambridge University Press.

Cantarero, D. (2006) Health care and patients' migration across Spanish regions. *European Journal of Health Economics, 7*, 114–116.

Cookson, R., Laudicella, M., & Li Donni, P. (2013). Does hospital competition harm equity? Evidence from the English National Health Service. *Journal of Health Economics, 32*, 410–422.

Cooper, Z., Gibbons, S., Jones, S., & McGuire, A. (2011). Does hospital competition save lives? Evidence from the English NHS patient choice reforms. *The Economic Journal, 121*(554), F228–F260.

Fabbri, D., & Robone, S. (2010). The geography of hospital admission in a national health service with patient choice. *Health Economics, 19*, 1029–1047.

Gaynor, M., Moreno-Serra, R., & Propper, C. (2010). Death by market power: Reform, competition, and patient outcomes in the National Health Service. NBER Working Paper 16164.

Gaynor, M., & Town, R.J. (2012). Competition in Health Care Markets. In T. Mc Guire M. Pauly & P. P. Barros (Eds.), *Handbook of Health Economics* (Vol. 2, pp. 499–637). Amsterdam: Elsevier.

Griffith, D. A., & Jones, K. G. (1980). Explorations into the relationship between spatial structure and spatial interaction. *Environment and Planning A, 12*, 187–201.

Le Sage, J. P., & Pace, R. K. (2008). Spatial econometric modeling of origin-destination flows. *Journal of Regional Science, 48*, 941–967.

Le Sage, J. P., & Pace, R. K. (2009). *Introduction to spatial econometrics*. Boca Raton: CRC.

Levaggi, R., & Menoncin, F. (2008). Fiscal federalism, patient mobility and soft budget constraint in italy. *Politica economica, 26*, 367–386.

Levaggi, R., & Menoncin, F. (2012). Soft budget constraints in health care: Evidence from Italy. *European Journal of Health Economics*. doi:10.1007/s10198-012-0417-4.

Levaggi, R., & Zanola, R. (2004) Patients' migration across regions: The case of Italy. *Applied Economics, 36*, 1751–1757.

Lippi Bruni, M., Nobilio, L., & Ugolini, C. (2008). The analysis of a cardiological network in a regulated setting: A spatial interaction approach. *Health Economics, 17*, 221–233.

Morrill, R. L., & Earickson, R. J. (1968). Hospital variation and patient travel distance. *Inquiry, 5*, 26–34.

Propper, C. (2012) Competition, incentives and the English NHS. *Health Economics, 21*, 33–40.

Rademakers, J., Delnoij, D. M., & de Boer, D. (2011). Structure, process or outcome: Which contributes most to patients' overall assessment of healthcare quality? *BMJ Quality and Safety, 20*(4), 326–331.

Victoor, A., Delnoij, D. M., Friele, R. D., & Rademakers, J. J. (2012). Determinants of patient choice of healthcare providers: A scoping review. *BMC Health Service. Research., 12*, 272.

The Impact of Federalism on the Healthcare System in Terms of Efficiency, Equity, and Cost Containment: The Case of Switzerland

Luca Crivelli and Paola Salari

Abstract According to the economic theory of federalism (Oates 1999), a decentralized decision to collectively fund and supply the quantity and quality of public services will increase economic welfare as long as three conditions are fulfilled: preferences and production costs of the different local constituencies are heterogeneous; local governments are better informed than the central agency because of their proximity to the citizens; and the competition between local governments exerts a significant impact on the performance of the local administration and on the ability of public agencies to implement policy innovation. Federalism also presents some negative aspects, including the opportunity costs of decentralization, which materialize in terms of unexploited economies of scale; the emergence of spillover effects among jurisdictions; and the risk of cost-shifting exercises from one layer of the government to the other. Finally, competition between fiscal regimes can affect the level of equity. The literature considers fiscal federalism as a mechanism for controlling the size of the public sector and for constraining the development of redistributive measures. The present paper reviews the impact that federalism has on the efficiency, equity, and cost containment of the healthcare system in Switzerland, a country with a strongly decentralized political system that is based on federalism and the institutions of direct democracy, a liberal economic culture, and a well-developed tradition of mutualism and social security (generous social expenditure and welfare system). By analyzing the empirical evidence available for Switzerland, we expect to draw some general policy lessons that might also be useful for other countries.

L. Crivelli (✉) · P. Salari
Department of Economics, Università della Svizzera Italiana (USI), Via Buffi 13 6900 Lugano, Switzerland
e-mail: luca.crivelli@usi.ch

P. Salari
e-mail: paola.salari@usi.ch

R. Levaggi and M. Montefiori (eds.), *Health Care Provision and Patient Mobility*, Developments in Health Economics and Public Policy 12,
DOI: 10.1007/978-88-470-5480-6_7, © Springer-Verlag Italia 2014

Keywords Federalism · Preference matching · Cost efficiency · Equity

JEL Classification D24 · D61 · D63 · H77

1 Introduction

The Swiss healthcare system is based on two distinct pillars. The first is a health insurance model that has atypical characteristics and lies somewhere between private and social insurance. This model was influenced by the country's culture and history and was recently reformed following the ideas of Alan Enthoven. The second element is a high degree of decentralization (since 1848 the federal Constitution has entrusted the country's 26 cantons with a large amount of autonomy in the organization of their regional health care sector).[1]

The interaction of these two factors gives rise to the observable health system outcomes in terms of equity and efficiency. However, it is difficult to determine the extent to which the general equity and efficiency of the Swiss health system depend on the particular health insurance model or reflect the high level of decentralization. For this reason, based on the available evidence in the literature, we will consider the heterogeneity (in terms of efficiency and equity) across the Swiss cantons as a proxy for the specific effect of decentralization.

The aim of this chapter is to provide an overview and discussion of the equity and efficiency outcomes that Switzerland's various cantonal systems have reached, relying on the extant literature. Section 2 introduces the general economic theory behind decentralization (that is, fiscal federalism), highlighting the possible advantages and drawbacks of such a system. The other sections focus specifically on Switzerland. Section 3 describes the main features of the Swiss healthcare system, its overall organization, and its historical background. Sections 4 and 5 discuss the implications of decentralization with regard to efficiency and equity outcomes, respectively, with emphasis on the differences among cantons. Section 6 concludes.

2 General Theory of Fiscal Federalism: A Short Literature Review

During the last century, many OECD countries have had a general tendency towards decentralization of governments. The concept of decentralization can have a number of different meanings. In this chapter, it refers to a setting in which the

[1] The precept of Swiss decentralization is that public policies and their implementation should be assigned to the lowest level of government that is capable of achieving the objectives.

central government has devolved some forms of autonomy (fiscal or organizational competences) to subnational governments (such as regions, or municipalities). The latter have the power to organize the provision of the public (and merit) goods that are mainly consumed in their territory. The idea behind decentralization is that it makes local governments better informed about the preferences of their citizens; so they can adapt public services to them.

2.1 The Theoretical Economic Background: Advantages of Decentralization

The principle of decentralization of competences and political power was inspired by the theory of fiscal federalism. In particular, the rationale stems from the "decentralization theorem" elaborated by Oates (1972). This theory starts with the hypothesis that each level of government has the goal of maximizing the social welfare of its area of competence. Therefore, decentralization is considered as a proper means for reaching such a goal. Subnational levels of government should provide "local" public goods; that is, goods that are primarily consumed in that territory. The decentralization theorem states that under certain conditions (rather homogeneous local preferences and rather heterogeneous preferences among subnational authorities), a vector of local public goods tailored to local preferences is expected to be Pareto superior to a situation with a uniform level of public output managed at the central level. There are at least two reasons to expect such a result. Firstly, decentralizing certain public services (such as healthcare) at a regional level makes it possible to rely on a regional preference matching (Oates 1972). Oates argued that the efficient level of a certain output may differ among jurisdictions because of different preferences. A subnational level of government is closer to the preferences of its citizens and can tailor the outputs of these goods and services to the preferences of the local constituency. This theorem holds true if we assume the presence of asymmetric information; that is, the local government knows the preferences of its citizens, whereas the central government does not. The second argument in favor of decentralization states that central governments have certain political constraints that prevent them from allocating different outputs to different areas. If different jurisdictions require a different Pareto-efficient level of such an outcome, then a central government cannot guarantee that this optimal level is reached (Oates 2005).

Federal organization of the state may stimulate horizontal competition among jurisdictions. Regions can compete to cater for the preferences of their own citizens in a "yardstick environment". Each region will offer a given quality and quantity of local public services according to citizens' preferences (and according to region's budgetary limit).

If we assume mobility of people, citizens in a federalist system can move to another jurisdiction that has more favorable conditions in terms of services, taxation, and so on. In his famous 1956 paper, Tiebout expressed the idea that people can "vote with their feet" (Tiebout 1956), meaning that people may move from one jurisdiction to another if they find the relationship between the supply of goods and services and the tax burden of the other jurisdiction more convenient.

A third argument in favor of decentralization is that a federal system promotes innovation in both the supply of public goods and in taxation policy. On one hand, it is easier to innovate in a small jurisdiction with a limited number of people than at a central level. The positive or negative effects of a new policy may be better controlled in a small territory and it is also easier to revert to the initial situation in case of failure. On the other hand, the federal setting may be conceived as a laboratory in which several solutions to the same problem can be tested against one another simultaneously. This can become a virtuous environment if local governments are willing to compare the various solutions and apply the policy with the best performance in their own jurisdiction.

All of the motivations described above can be summarized in the following simple concept: decentralization helps to achieve better results in terms of allocative efficiency, due both to regional preference matching that better expresses citizens' tastes and horizontal competition among regions.

In the last decade, a so-called "second-generation theory" of fiscal federalism has been developed. This strand of literature has a wider perspective on federalism and combines economic and political views of the problem. This theory is focused on the behavior of political agents in a context of imperfect information. The second-generation theory shifts the focus towards the behavior of the agents, noting the importance of the information that all the agents involved in the process possess. In this vision, federalism is seen as the result of political bargaining between politicians and voters (see Oates 2005).

This more recent strand of research has highlighted another positive aspect of a federal organization. At a local level, politicians may have higher "electoral accountability" towards their electors, which should lead to the incumbent having a higher sense of responsibility and commitment.

In the last century, several countries have encouraged decentralization in the organization of the healthcare sector. Even though fiscal federalism was not designed for the provision of private merit goods such as healthcare, there are a number of reasons why healthcare is a good field where decentralization should be implemented. In this sector, the preferences of the consumers (the patients) can be similar within the same region and different between regions (depending, for example, on the share of elderly people present in a specific area). At the same time, the healthcare supply can have many differences between regions (such as the density of physicians in the population, the topographic conformation of a territory, etc.) that require a different organization of the system. Therefore, a decentralized organization may be more responsive to people's needs and more efficient at adapting the specific supply-side structure. Nevertheless, a decentralized system does not always outperform a centralized one. Federalism also presents certain drawbacks that might yield both equity and efficiency concerns.

2.2 Challenges of a Decentralized System with Respect to Equity

One serious drawback of federalism is the possible increased heterogeneity in the services delivered across regions, which can go so far as to put the principle of social citizenship at risk (Banting and Corbett 2002). By definition, federalism cannot guarantee the principle of horizontal equity across a whole country since sub-central governments may choose different tax rates and the organization of public service provision might differ (in terms of quantity, quality and prices) according to the preferences of the different populations. Consequently, strong disparities in the treatment of citizens who live in different areas of a country are immanent in federal states. These disparities may have important socioeconomic effects in contexts such as healthcare, where differences in service organization may produce undesirable differentials in terms of quality of care and access to important health services. This effect may be even more pronounced if regions have very different economic situations. The risk is that the provision of public services will be driven by different levels of available resources rather than by the different preferences of the citizens.

To counter these equity problems, federal governments generally use two different strategies, which are often jointly applied. The first strategy is to define in a federal policy framework the minimum standard level of services that should be guaranteed in all jurisdictions. The second strategy is to rely on equalization mechanisms based on vertical (and in some cases also horizontal) financial transfers.

Through the equalization mechanism, the central state redistributes resources to reduce economic inequalities between regions and to correct for possible fiscal imbalances until a minimum service level (fixed in advance) is reached. The so-called intergovernmental grants may be conditional (such as matching grants) or unconditional (for example, lump-sum grants) and may have at least the three following functions (Oates 1999): they internalize the possible spillover effects of one jurisdiction to the others; they guarantee fiscal equalization across jurisdictions; they promote a more equitable tax system. In practice, intergovernmental grants can be seen as a means to redistribute wealth from the richer regions to the poorer ones so that each region may offer a similar level of services. In some regions, the provision of public goods may actually be smaller, not due to differences in preferences, but to differences in resources; this raises equity concerns.

Transfers are generally assigned according to a specific formula based on available resources (fiscal capacity) and (fiscal) needs of subnational governments. In some cases, there is a specific formula for a sector like health, while in other cases health is included together with other public functions in a general transfer.

Transfers are the most important instrument that governments have to redistribute resources and to reduce inequity issues. They are justified on the grounds of equity principles, in that they allow fiscally weaker jurisdictions to compete with stronger ones (Oates 1999). Transfers are also an insurance against specific shocks, such as epidemics (Costa-Font 2012). Nevertheless, there are also some negative aspects of transfers, such as the possibility of moral hazard problems. This phenomenon is implicit in the so-called "flypaper effect", whereby unconditional grants have a greater effect in increasing regional spending than a general rise in community income. This can be seen as a form of moral hazard.

Apart from vertical transfers, another solidarity instrument relies on certain risk adjustment principles among regions. In order to provide mutual insurance against risks related to health and healthcare, regions adjust for the differences in risk among specific patient groups through a compensation system. This may be seen as a form of horizontal transfer.

In the case of healthcare, however, a federal state can also offer another option to mitigate possible equity problems. This option involves giving citizens the option to be treated in another district where the quality of care is higher. This phenomenon is considered a less radical way of voting with one's feet. Although in this case the person does not move her residence to another region (as it would be the case in the original Tiebout's situation), she might decide to use the healthcare supplied in a region that provides a combination of quality and quantity of that service which better reflects her preferences. This may be possible only if the patient's mobility is permitted under conditions defined by the central government. Levaggi and Zanola (2004) and Crivelli (1998) have provided evidence of internal patient mobility in presence of differences in the quality of services across Italian regions and Swiss cantons, respectively.

2.3 Challenges of a Decentralized System in terms of Efficiency

A decentralization process also presents some other shortcomings related to efficiency.

Firstly, local governments may be inefficient due to the failure to internalize inter-jurisdictional externalities; spillover effects are difficult to include in the costs of the service. Such a problem occurs when some people in a jurisdiction free-ride by enjoying positive effects of a service provided by another jurisdiction without paying any correspondent tax for it. From the point of view of citizens of the jurisdiction that provides the service, this is an oversupply of goods. Spillover effects are responsible for a distorted allocation of public goods. A common solution in a case in which the provision of a local service produces spillover effects for citizens of another jurisdiction is that the central government may allocate a Pigouvian subsidy (in the form of matching grant) to compensate for these effects.

Another problem arises as a direct consequence of tax competition among subnational levels of government. Competition may lead to substantial allocative distortions. In order to attract wealthy taxpayers, regions may compete by holding the taxation burden below the optimum level that would reflect voters' preferences. At the same time, regions may find it convenient to relinquish an egalitarian distribution of income; they might have an interest in designing a relatively low level of social assistance (such as subsidies and other social aids for the worse off) to avoid an inflow of needy people (Feld 2000). On these grounds, Musgrave (1971) and Brown and Oates (1987) theorized that the task of income redistribution should be entrusted to the central government. Shifting this responsibility to decentralized levels of government could lead to the design of public policies that encourage positive self-selection. This could be done by choosing a high level of public services preferred by the better-off (such as elegant streetlights and luxuriant public gardens) and a low level of those services that are more appreciated by the worse-off (canteens in public schools, for example). This phenomenon is also known as "race to the bottom" (Oates 1999), which indicates harmful competition that could lead to a low taxation level, low public expenditures, and to a suboptimal level of some public services.

Another drawback that Oates (1999) identified are unexploited scale economies, with resulting expenditure increases. The question of whether fiscal federalism increases or decreases total expenditures remains open to debate. Most of the success of decentralization depends on the starting point; that is, whether decentralization can embrace heterogeneity or not. Most important, however, are the incentives and the institutional design in which decentralization develops. Decentralization offers some efficiency advantages if it is able to exploit local heterogeneities and if these more than offset a possible loss in the exploitation of the existing economies of scale; otherwise it may lead to an increase in general costs.

Costa-Font (2012) argued that many of the possible negative side-effects of federalism refer to the sunk costs of a change from a centralized system to a decentralized one. For example, duplicity costs might arise at the beginning of this process due to a possible overlap of competences regarding specific functions or responsibilities. Such sunk costs are related to the need for coordination among different levels of government, which is especially costly at the beginning of the decentralization process. This would not apply in countries like Switzerland, where decentralization of competences has always existed in the organization of the healthcare system.

Another cost to consider is the one paid by the central government to eliminate the bailout expectations of the regions. Bordignon and Turati (2009) considered the role of bailing out expectations as an important determinant of public expenditure. Their study is based on a strand of literature that agrees on the importance for central government to commit not to bail out regional supplementary expenditure. They argued that if regions expect that citizens of the other regions will pay their bill, it is likely that moral hazard behavior will result, as local governments will have an incentive to waste money and inflate their expenditures. This important issue is known as "soft budget constraint", because the local governments' budget constraint may become "soft" and lead to an increase in the level of expenditures. Changing these expectations may help contain public expenditure.

Finally, as many authors have highlighted (e.g., Costa-Font 2012), another obstacle to the success of a decentralized system is that the central government must devolve responsibilities not only on the side of expenditures, but also on the side of taxation. If the autonomy is allowed only on the expenditures side, there is a risk of expenditure expanding, which presents obvious problems in terms of efficiency.

3 The Swiss Health System

Switzerland is a small federal state (7.9 million inhabitants as of 2012) made up of 26 cantons. The present organization of the Swiss health sector reflects at least three fundamental elements (Achtermann and Berset 2006): (a) a strongly decentralized political system, based on federalism, subsidiarity, and the institutions of direct democracy; (b) a liberal economic culture that emphasizes freedom of choice and consumer-driven economic decisions; and (c) a unique historical path for social security, in which non-profit institutions in the nineteenth century led to the creation of a voluntary insurance sector and have been influencing the design of universal coverage in Switzerland until the present day.

The Swiss health insurance sector was heavily reformed in 1996 according to Enthoven's principles of regulated competition (Enthoven 1993). Since then, Switzerland has had a federally established universal health insurance system with atypical characteristics lying somewhere between private and social insurance (Leu et al. 2007; OECD/WHO 2006) and with competition playing out within a national regulatory framework, but mostly at the cantonal (decentralized) level. Federalism[2] and the institutions of direct democracy,[3] along with the pre-existence

---

[2] Article 3 of the current Swiss Constitution establishes the high degree of autonomy to the cantons, stating that 'The cantons are sovereign insofar as their sovereignty is not limited by the Federal Constitution; they shall exercise all rights which are not transferred to the Confederation.'.

[3] In the Swiss political system (both at the cantonal and federal levels) citizens have the opportunity to participate directly in every state decision by means of direct democracy. For example, federal laws and generally binding decisions of the Confederation are subject to an optional referendum; in this case, a popular ballot is held if 50,000 citizens request it. The referendum is similar to a veto and has the effect of delaying and safeguarding the political process by blocking amendments adopted by parliament or the government or delaying their effect. Accordingly, referenda are often described as a 'brake' applied by the people. A second way for citizens to induce a change is called 'popular initiative'. If at least 100,000 signatures are collected within 18 months to propose a constitutional amendment, then a popular ballot must be held. The outcome will be binding, as long as a majority of voters and cantons support the proposal.

of a large number of mutual support groups,[4] have strongly conditioned the fundamental choices of the Swiss welfare system (Gilliand 1986). In 1899, the Swiss federal assembly envisaged setting up a system of public funds inspired by the Bismarck model. This system would have been jointly financed by the insured and employers and organized on a territorial basis starting with a minimum number of 1,500 insured (Gilliand 1990). It was undoubtedly a more modern and rational system of health insurance than the highly fragmented one that had spontaneously emerged in the nineteenth century. However, in May 1900 the reform was rejected in a referendum. From the ashes of that ballot, the first federal Law on Sickness and Accident Insurance developed a decade later. The law was accepted by parliament in 1911 and approved by the people in the referendum held in 1912. The legislature realized that to overcome the obstacle of direct democracy and introduce a federal law on the subject of health insurance, it was necessary to leave the management of the sector in the hands of private institutions without restraining cantonal autonomy.

Unlike the Bismarck model, the law of 1911 saw the Swiss legislature relinquish the idea of making health insurance compulsory on a national scale; instead, it left it up to the cantons to decide whether to make it compulsory at the cantonal level.[5] Insurance premiums were established by each sickness fund and were not related to income but were adjusted for age and gender. To reduce the financial fragility of the sickness funds and to stimulate voluntary affiliation, the Confederation decided to participate in the financing of premiums with public money by transferring a lump-sum per capita subsidy to the sickness funds. The earliest available statistics show that approximately half of the Swiss population was insured in 1945, while nearly universal coverage was achieved between 1985 and 1990.

The organization of healthcare delivery, particularly that of inpatient care, has historically come under the control of the cantons (Kocher and Oggier 2007). Moreover, the decisional autonomy of cantons has been combined with decentralized financial responsibility (Wyss and Lorenz 2000). The public expenditure of cantons and municipalities in Switzerland represents approximately 70 % of total public expenditure, while only 30 % is paid by the Confederation. In terms of direct public spending in the healthcare sector, the part funded by the federal government is only approximately 2 % (Gerritzen and Kirchgässner 2013). In order to guarantee an equitable amount of resources, Switzerland has implemented a fiscal equalization system based on vertical and horizontal grants, with the aim of decreasing differences between rich and poor jurisdictions.

---

[4] A census held in 1903 counted 2,006 mutual support groups, to which 14 % of the population was affiliated (approximately 500,000 people). Half of the groups had fewer than 100 members and grouped together the inhabitants of one municipality.

[5] Prior to 1994, six cantons made affiliation to a sickness fund compulsory for the whole population; 12 cantons made affiliation com-pulsory for certain social groups such as people with a low income and foreigners; and four cantons delegated the decision for a man-date to each municipality (Alber and Bernardi-Schenkluhn 1992: 210).

Despite fiscal equalization, strong decentralization of competences, ample autonomy of the cantonal governments in public expenditure decisions, and fiscal federalism has, over time, created significant differences among cantons with respect to per capita healthcare spending, regulatory setting, the role of the private versus public sector, and the level of production capacity (Crivelli et al. 2006; Vatter and Rüefli 2003). In addition, federalism has encouraged the proliferation of organizational models that vary significantly across cantons. Finally, direct democracy and federalism were at the origins of the very slow pace of radical reforms at the national level. Referenda and popular initiatives have allowed Swiss citizens to intervene directly in the decision-making process, approving or rejecting each reform via a popular ballot. Because unbalanced and radical revisions have a high likelihood of rejection in popular ballots, bills are generally amended early in a pre-parliamentary phase that involves negotiation ex ante with opponents of reforms originating in government or parliament and incorporating the demands of the most powerful lobbies (Cheng 2010). Between 1974 and 2012, the Swiss population was called to the ballot box 11 times to deliberate on reforms in the health insurance sector.[6] With the exception of the referendum on the Federal Health Insurance Act (KVG), which was approved in 1994[7] and came into force in January 1996, the remaining 10 popular ballots failed. It took 85 years to profit from a "windows of opportunity" and to pass a radical system reform (Crivelli 2014).

Today, the Swiss healthcare sector is still composed of 26 cantonal systems, connected to each other by the KVG. However, while each canton is still formally responsible for ensuring access to good quality health services, the KVG has shifted the balance of power from the cantons to the Confederation. Health insurance is now compulsory at the federal level and the Confederation defines the benefit package that is guaranteed to each resident[8] and financed by two main instruments: compulsory insurance supplied by private sickness funds within the framework of the KVG and the public spending of the cantonal and municipal authorities. The latter is financed by general local government taxation and used to subsidize providers who offer services included in the compulsory benefit package (such as hospitals, nursing homes, and public and non-profit home care institutions). Furthermore, both federal and cantonal contributions are used to subsidize health insurance premiums for households with modest incomes.

The KVG imposed a reduction in cantonal autonomy on decisions regarding public expenditure, which led to several important changes in the distribution of

[6] Three reforms of the federal law proposed by parliament and put to referendum, six popular initiatives, and two counter-projects.

[7] A narrow majority of 51.8 % voted in favor of the new law.

[8] Social insurance is not automatic, but it is compulsory. The cantons are responsible for the surveillance of this mandatory insurance and checking the membership status of each citizen. It is impossible to leave one sickness fund without having a contract with another insurer and fines are imposed on those who are caught without coverage (Brunner et al. 2007: 151–2; Cheng 2010: 1443).

tasks between the Confederation and cantons.[9] In the future, the Confederation is expected to play an even more important role in defining Swiss health policy,[10] with a view to thwarting regional differences in supply, harnessing economies of scale, and curbing the growth of health expenditure at national level. However, the transfer of new tasks to the Confederation cannot take place effectively without an amendment to the Federal Constitution (Schaffhauser et al. 2006) and must be accompanied by a corresponding adjustment of the public expenditure share borne by the federal government (Crivelli and Filippini 2003). This missing constitutional reform is a violation of the principle of 'who decides, pays', since the bulk of public spending on health is still financed by the cantons, even though the Confederation plays an increasingly important role in health policy decisions. Accordingly, it is not surprising that, in the last decade, cantons have been unwilling to accept radical reforms of the system aimed at transferring additional responsibilities and decision-making power to the central government and to health insurers without an equivalent transfer of financial responsibilities. Cantons are currently the main opponents of the federal government's roadmap of reform, which has made the search for consensus on fundamental changes slow and complex (Crivelli and Bolgiani 2009).

The following sections will discuss in greater detail the aspects of equity and efficiency that characterized the Swiss health system. These are two sides of the same coin that need to be viewed together.

4 Fiscal Federalism and Efficiency in the Swiss Healthcare System

As explained in Sect. 2, the decentralization of the organization and production of healthcare has the advantage of tailoring the supply to the preferences and specific needs of the population. In Switzerland, the heterogeneity of preferences across the constituencies of the three main linguistic areas[11] and/or of the rural versus urban cantons is highly significant and documented. The heterogeneity emerges in examples such as the level of acceptance for managed care contracts (see Zweifel et al. 2006 who used discrete choice experiments to account for differences in willingness-to-pay between the German and French speaking cantons), in the attitudes towards risk (such as the choice of higher deductibles) and in voters' preferences expressed in popular ballots (Crivelli 2014). This heterogeneity

---

[9] Switzerland is moving in the direction hoped for by the theory of fiscal federalism, according to which the central government should have responsibility for income redistribution (and therefore also for financing the basic stock of merit goods), whereas cantons should be responsible for the organization and production of health services (Oates 1999).

[10] In 2013, for the first time, the federal government issued a national health care strategy (see Federal Office of Public Health 2013).

[11] Switzerland has a German-speaking, a French-speaking and an Italian-speaking part.

suggests that, in Switzerland, federalism could lead to a better preference matching and to the achievement of higher allocative efficiency compared to a centralized ("one size fits all") solution. However, the success of decentralization also depends on the dimension of the local jurisdictions and on the coordination between levels of government.

With 26 jurisdictions (cantons) in Switzerland for a population of just 8 million, at least two other efficiency-related aspects must be considered. Firstly, an important goal of every productive system (including decentralized ones) is to reach technical efficiency (that is, to produce the maximum quantity of outcome with the minimum level of inputs, by exploiting the available economies of scale). Secondly, in a decentralized setting with a multi-level government organization, there is a higher risk of cost shifting (Banting and Corbett 2002). This cost shifting could occur if several public payers (in addition to private and social insurers) are jointly responsible for healthcare funding and lack the incentives to contain total health expenditure. In fact, total cost containment might be more difficult to achieve (or may imply higher political costs) than containment of the individual financial burden since, in the presence of multiple payers, a viable option could be to shift part of the expenditure to another payer. This section discusses both aspects in detail.

As Costa-Font (2012) highlighted, there is no clear evidence about the impact of decentralization on the level of expenditure. It is still unclear whether decentralization reduces the level of expenditure due to an increase in the efficiency of the entire system or yields a higher level of activities that requires higher expenditure in the end. The answer to this question depends on many factors, such as the design of fiscal policy and fiscal imbalances, policy innovations, the promotion of competition, etc. In short, it depends on the incentives embedded in the used sources of financing.

It is not always simple for a decentralized system to guarantee technical efficiency. A potential problem that may occur in the presence of fiscal federalism is the suboptimal dimension of care institutions (such as hospitals and nursing homes) that might not entirely exploit the existing economies of scale. In a decentralized system, regional governments may be responsible for hospital planning and municipalities of nursing homes administration, although it would be more economically and socially efficient to assign the control of these areas to larger jurisdictions. Jurisdictions are sometimes too small to exploit economies of scale (Eichenberger and Frey 2006). This is common in Switzerland, where hospitals are managed at the cantonal level and nursing homes at the municipal level. Several studies have found that the vast majority of Swiss nursing homes and hospitals do not reach the optimal dimensions (see, e.g., Crivelli et al. 2001, 2002; Farsi and Filippini 2006). Therefore, it would be possible to have an efficiency gain if these structures were designed and managed by larger jurisdictions. In theory, small cantons (or municipalities) could build consortia to jointly run hospitals (or nursing homes). Unfortunately, this solution is more appealing in theory than in practice, since it leads to artificial political entities without the necessary electoral accountability to function properly and facing a lack of fiscal equivalence.

Widmer and Zweifel (2012) showed how, in the case of Switzerland, the wrong incentives that are immanent in every fiscal equalization program led to low performance in terms of efficiency. They argued that cantons with higher financial potential (that is, the net-payers of the equalization system) may have incentives to underperform. Moreover, earmarked subsidies may encourage inefficiency. To test this hypothesis, Widmer and Zweifel (2012) performed a data envelopment analysis at the cantonal level, which enabled them to assign an efficiency score to each canton for six categories of services considered, including healthcare. They then checked for any correlation between the scores obtained and a vector of variables representing the Swiss fiscal equalization program.

Their results confirmed the hypothesis of a negative correlation between the amount of financial equalization and cantonal efficiency, meaning that the equity-efficiency trade-off advocated by Stiglitz (1988) does exist in Switzerland. This is more evident in payer cantons than in the receiver ones. Widmer and Zweifel argued that cantons that do not expect to receive anything from the equalization will prefer to waste their resources instead of giving them to the worse-off cantons. They also found a negative correlation between earmarked federal subsidies and cantonal efficiency: cantons receiving a federal subsidy are more likely to be inefficient. Finally, they found that the best efficiency scores are reached by small rural cantons rather than by urban ones. Nevertheless, the inefficiency scores that the authors used in their correlation analysis are an average of the indices found in all the categories, not just health.

Another aspect strictly related to the decentralized structure of the health system is the evolution of the health expenditure, which could be due to the adoption of cost shifting strategies. In 2010, the total health expenditure in Switzerland was approximately 10.9 % of GDP, which is one of the highest rates in the world.

Moreover, per capita health expenditure in Switzerland shows conspicuous variation across cantons. The first column of Table 1 highlights the level of per capita socialized health expenditure (SHE), which ranges from 3579 CHF in Thurgau to almost twice that (6908 CHF) in Geneva.[12] It is difficult to determine how much of this heterogeneity reflects cantonal preferences and how much is due to better cost containment or to a more rational organization of the health service supply in the cantons with lower spending. Crivelli et al. (2006) empirically assessed the determinants of the cantonal health expenditure using panel data analysis. They found that these differences depend on economic, demographic, and also structural factors.

Crivelli et al.'s analysis also shows that an increase in the physician density determines, *coeteris paribus* an increase in the socialized health expenditure. This

[12] Socialized health expenditure (SHE) reflects collective spending for the universally accessible basket of healthcare benefits. This includes public financing, mandatory health insurance and social insurance and accounts for approximately 60–65 % of the total financing in Switzerland. Out-of pocket expenditures, co-payments and voluntary health insurance are not included.

Table 1 Socialized health expenditure (absolute value and share of each financing source), amenable mortality, and Kakwani index (averages 2004–2008)

Cantons	SHE (CHF per capita 2004–2008)	Amenable mortality (age-standardized rates per 100,000 population)	Sources of financing (2004–2008)				Kakwani Index (1998–2005)
			Federal taxes (%)	Cantonal and municipal taxes (%)	Non health social insurances (%)	Social health insurance (net premiums) (%)	
Zurich (ZH)	4396	57.2	5.90	33.00	14.30	46.80	−0.108
Bern (BE)	4829	54.3	8.50	32.60	13.00	45.90	−0.095
Luzern (LU)	3816	55.6	9.30	29.70	16.50	44.40	−0.094
Schwyz (SZ)	3762	60.1	6.80	29.30	16.70	47.10	−0.134
Obwalden (OW)	3666	56.6	12.60	29.90	17.20	40.30	−0.057
Nidwalden (NW)	3664	46.4	6.00	33.30	17.20	43.50	−0.073
Glarus (GL)	3860	64	8.00	29.40	16.30	46.30	−0.091
Zug (ZG)	4103	51.7	4.80	38.10	15.40	41.70	−0.087
Fribourg (FR)	4333	56.2	9.60	32.60	14.50	43.30	−0.087
Solothurn (SO)	4324	62.9	6.90	30.50	14.60	48.00	−0.087
Basel-Stadt (BS)	6370	60.4	4.70	43.90	9.90	41.50	−0.058
Basel-Landschaft (BL)	4414	51.3	6.10	29.40	14.30	50.20	−0.077
Schaffhausen (SH)	4409	67.6	7.70	34.30	14.30	43.70	−0.154
St. Gallen (SG)	3772	58.1	7.10	29.10	16.70	47.10	−0.107
Graubünden (GR)	4124	54.7	7.40	32.50	15.30	44.80	−0.129
Aargau (AG)	3772	55.8	6.10	24.80	16.70	52.30	−0.107
Thurgau (TG)	3579	56.9	9.70	24.60	17.60	48.10	−0.113

(continued)

Table 1 (continued)

Cantons	SHE (CHF per capita 2004–2008)	Amenable mortality (age-standardized rates per 100,000 population)	Sources of financing (2004–2008)				Kakwani Index (1998–2005)
			Federal taxes (%)	Cantonal and municipal taxes (%)	Non health social insurances (%)	Social health insurance (net premiums) (%)	
Ticino (TI)	5035	50.5	7.50	34.60	12.50	45.40	−0.059
Vaud (VD)	5176	49.7	6.90	34.80	12.20	46.10	−0.100
Wallis (VS)	4213	51.7	10.60	32.50	15.00	42.00	−0.111
Neuchâtel (NE)	5196	53.3	8.00	36.00	12.10	43.90	−0.09
Geneva (GE)	6908	44.9	4.60	46.60	9.10	39.70	−0.064
Jura (JU)	4613	51.9	9.80	32.10	13.70	44.50	−0.057
Switzerland	4556	54.7	7.20	33.40	13.80	45.60	−0.100

Canton Uri, canton Appenzell Inner-Rhodes, and canton Appenzell Outher-Rhodes have not been considered because of a lack of data may reveal the existence of a supply-induced demand effect: the presence of more doctors in the territory induces a higher use of healthcare, since providers in Switzerland are still paid according to a fee-for-service reimbursement scheme. Therefore, cantons seem to differ not only in terms of demographic conditions and citizens preferences, but also in their regulatory efforts and ability to keep the medical supply under control.

A more recent study (Reich et al. 2012) includes some additional variables and uses a longer panel, but significantly confirms the previous results. The panel data analysis provides evidence of a strong correlation between expenditure and the per-capita rate of specialist physicians and dispensing doctors,[13] income, the share of managed care contracts and other socio-economic factors. The main message of Reich et al.'s study is that a more integrated model in the provision of ambulatory care may curb health expenditure. By increasing the availability of integrated care networks and the acceptance of managed care plans in the population, per capita health expenditure may be better contained.

5 Fiscal Federalism and Equity in the Swiss Healthcare System

This section discusses how the federal setting of the Swiss healthcare system impacts on equity.

Although Switzerland usually relies on economic liberalism for most of its policy decisions, in the healthcare field it has embraced an egalitarian point of view of social justice and included equity as one of the main aims of the healthcare system (similarly to the majority of the OECD countries).

The features that an equitable healthcare system should guarantee may be summarized in two main concepts: horizontal and vertical equity. The former means that people in a similar situation (with similar needs, similar economic situation) have to be treated similarly. The latter expresses the idea that different people need to be treated differently; it is generally used to explain the principle of equity in financing, referring to the fact that people in different economic situations must contribute differently to the health system financing.

The strand of literature concerning the equity concept in health economics is based mostly on empirical analysis (O'Donnell et al. 2008) and generally concentrates on three main measures of equity: equity in health itself, equity in access to (or use of) healthcare services (according to the idea of "equal treatments for equal needs"), and equity of the healthcare system financing, which guarantees that the financial burden of healthcare is distributed according to the ability of individuals to pay (that is, progressively), which places greater financial responsibility on the better-off than on the poor (van Doorslaer et al. 1992).

[13] Doctors with legal authorization to sell pharmaceutical products directly to their patients.

This section concentrates primarily on the equity of financing measure, due to the lack of research at cantonal level in the two other fields. Therefore, this is the only equity dimension that allows us to draw some conclusions about the impact of decentralization on equity. As far as the equity of health status and the equity of access are concerned, we briefly review the overall situation of Switzerland in the next paragraph.

5.1 Equity of Health Status and Equity of Access

The ultimate outcome of any healthcare systems is generally recognized as the achievement of the maximum level of health for the population (WHO 2000). Accordingly, the equity in access to health services is considered instrumental to the goal of equity in the level of health. In order to guarantee equitable health status to citizens, public policies should restrain social gradients on health: a good level of health should be guaranteed to each citizen, regardless of his or her socio-economic condition. The literature on Switzerland highlights a weak positive relationship between family income and health status (Leu and Schellhorn 2004a; Reinhardt et al. 2012), which reveals the existence of a social gradient on health status. Nevertheless, the severity of this problem is less pronounced than it is in the rest of Europe. In fact, the magnitude of the Swiss social gradient is the second smallest in Europe[14] (Leu and Schellhorn 2004a).

Research into the equity of access to (or use of) health services intends to measure horizontal equity; that is, the extent to which people with similar needs are treated equally. An issue of inequity may arise if people who rely on better socioeconomic conditions have easier access to healthcare services.

The literature covering this topic has reached varying results. Considering all physicians visits, there is evidence of a slightly pro-poor distribution (van Doorslaer et al. 2000). If GP and specialist visits are considered separately, a slightly pro-poor distribution in the number of GP visits has been found for Switzerland, while specialist visits were pro-rich distributed (Leu and Schellhorn 2004b). However, Allin et al. (2009) found no robust evidence of any pro-rich inequalities with respect to specialist visits for a sample of elderly people.

Given the lack of literature on cantonal differences in this field, we have shifted our focus to an index that could be considered as a measure of service quality; namely, the age-standardized rate of amenable mortality per 100,000 inhabitants (for the definition of this concept, see Nolte and McKee 2011). Although this index cannot replace the results of a comprehensive equity analysis, it does indirectly contain information about the access to (and the quality of) the health services: a faster and easier access to (higher quality) health services may reduce the

[14] The Netherlands is the only country with lower income-related inequality in health status.

amenable mortality rate. This information enables us to make comparisons across cantons.

The second column of Table 1 reports the index for the Swiss cantons. The variation is not huge, but it is significant, suggesting the existence of important cantonal variations in the quality of the healthcare services. The gap between the lowest value of canton Geneva (44 amenable deaths for every 100,000 people) and the highest value of canton Schaffhausen (68 amenable deaths for every 100,000 people) is comparable to that detected between different countries such as France and Germany (Nolte and McKee 2011).

5.2 Equity of Financing

The concept of equitable financing suggests that the economic burden of healthcare should be distributed between individuals according to their ability to pay, regardless of their health status or utilization of health services. This implies not only a form of solidarity between the sick and the healthy, which is implicit in any health insurance system, but also solidarity between rich and poor.

The financing sources of the Swiss healthcare system are a combination of private and public. The main "financing actors" are households, which buy social and private health insurance and pay out-of-pocket care, the state, and businesses that have to fund general social insurance.[15] Approximately two-thirds of the financing is collected directly from households (independent of their ability to pay) through monthly premiums, deductible, copayment, and out-of-pocket expenses. Thus, this part of the financing is likely to be regressive and to give rise to equity issues. Only one-third of financing is likely to be proportional or even progressive, being collected according to households' ability to pay through general taxation and general social insurance.

Some of the financing strategy rules are decided at the federal level. However, Swiss federalism allows cantons to make their own decisions regarding certain aspects of the financing process. The most important choice regards earmarked subsidies; that is, the policy tool used to smooth the regressivity of health insurance premiums and to redistribute income from the rich to the worse-off.

Economic theories disagree about which level of government should be in charge of redistribution. Oates (1972) felt that because redistribution is a national public good, the central state should handle it. However, other economists see redistribution as a local public good (Pauly 1973) and argue that the subnational governments should perform it. The Swiss model is a mixed one: subsidies are jointly funded by the Confederation and cantons (with an almost 50/50 share).

[15] We refer to a Bismarckian insurance model, funded by means of wage-based social contributions and designed to cover, through in-cash benefits, other health-related risks like longevity, disability, and accident.

However, cantons may determine the rule through which allocate subsidies; they fix the amount of subsidies and also the income threshold level under which people are allowed to obtain them. Moreover, the federal setting allows each canton to differ also in its level of (cantonal and municipal) taxation. According to all these cantons' choices, different equity (or inequity) levels are determined.

Columns 3–6 of Table 1 list the share of each financing source in the socialized health expenditure. The average values for Switzerland suggest that 45 % of socialized health expenditure is funded through health insurance premiums; that is, the most regressive financing source. The financing coming from taxation represents roughly 40 % of the socialized health expenditures (33.4 % cantonal and municipal taxation and 7.2 % federal taxation). The smallest part of financing (13.8 %) is collected proportionally to income through a general social insurance that pays for treatments in case of accident or disability.

The table shows that the composition of socialized health expenditure differs quite substantially among cantons. Consider the two extreme cases of Aargau and Geneva: the former finances more than half of its SHE through net premiums (52 %), compared to 40 % for the latter. On the contrary, the share of the cantonal and municipal taxes on total expenditure in Geneva (46.6 %) is almost twice as much as in Aargau (25 %).

Nevertheless, this data does not provide precise information about the extent of regressivity of the financing mix. In order to analyze this aspect, we need to rely on the Kakwani index, the most commonly used indicator for this kind of analysis. Wagstaff et al. (1999) estimated the equity of healthcare financing for Switzerland and other 12 OECD countries through the Kakwani index based on data from the early 1990s. This index, which is generally used in the public economics realm to check the progressivity of taxation, measures the progressivity (or regressivity) of a tax in terms of shifting from proportionality. A positive value of the Kakwani index indicates that the tax (in our case the financing-mix) is progressive (that is, the financial contribution of the rich is larger not only in absolute terms, but also as a percentage of the disposable income if compared with the poor), while a negative value suggests that the financing source in question is regressively financed. The computed Kakwani index for Switzerland (−0.1402) was negative, meaning that poor people financed the system relatively more than rich people.

Bilger (2008) used a different technique to check the level of regressivity of the Swiss health system, based on more recent data (from 1998, after the KVG reform). His results confirmed those from elsewhere; namely, that the Swiss health system is still very regressive (also after the major reform of 1996).

Crivelli and Salari (2012) analyzed the equity of the system exploiting the federal dimension of the Swiss system. The Kakwani index was computed using household data for the period from 1998 to 2005 to assess the level of regressivity in each canton for each financing source separately. These partial results were then weighted using the share of each financing source in the cantonal SHE to compute the overall index. The results suggest that the financing mix is highly regressive, especially in the cantons where the total level of health expenditure is smaller. The most regressive financing source turns out to be the net premiums.

The final column of Table 1 shows the value of the total Kakwani index in 23 out of 26 cantons. The heterogeneity in the results is evident: the difference between the lowest and the highest Kakwani index is (in absolute terms) comparable to that between USA and Sweden (as computed by Wagstaff et al. 1999), two countries with completely different healthcare systems. This large heterogeneity depends on the choices made by cantons in terms of subsidies (the amount and income level under which people are eligible to receive them), but also on the amount of public financing they decided to allocate, the progressivity of the cantonal tax system, and many other decisions (for more details, see Crivelli and Salari 2012).

As expected, cantons with a more generous policy in terms of subsidies, and with a widener public financing with respect to the Swiss average are those that perform better in terms of equity. For example, Jura and Geneva have a smaller Kakwani index than the other cantons and are among the most generous in terms of subsidies, considering both the share of population who receive them and the average amount of the subsidy (OECD/WHO 2011).

However the general finding suggested by this research is that even in the most generous cantons, the subsidy policy designed to smooth the regressivity of health insurance premiums is not large enough to turn the financing mix into a proportional payment; the overall SHE financing is regressive everywhere. To check for a possible trade-off between equity and efficiency, these results have been ranked and compared through a Spearman test and a Kendall test to rank their correlation to the efficiency scores found in Widmer and Zweifel (2012). Interestingly, the hypothesis of independence is not rejected, meaning that, in this particular setting, there is no evidence of any direct links between equity and efficiency.

Moreover, from a federal perspective, there is an additional notable aspect. On the grounds of empirical evidence, a 'race to the bottom' may yield worse results in terms of vertical equity than would be the case in a centralized state. This seems to contradict the theory that the Swiss health system setting—that is, a combination of community rating and premium subsidies—might be the most equitable solution (Kifmann and Roeder 2011).

Interestingly, Gilardi and Füglister (2008) found that cantons are generally more likely to imitate neighboring cantons in terms of subsidy policy: this may provide additional evidence of a possible 'race to the bottom'. Cantons are likely to maintain a lower level of subsidies to avoid an inflow of poor people from the neighboring cantons, lowering the overall equity of the system.

6 Conclusions

This chapter has presented an overview of the differences in equity and efficiency across the Swiss cantonal healthcare systems, linking these results to fiscal federalism and to the substantial decentralization of healthcare in Switzerland since the federal Constitution of 1848.

We started by describing the main features of decentralization according to the economic theory of fiscal federalism. Such an organizational model can have many economic advantages. Subnational governments are supposed to be closer to the preferences of the citizens, whereas a yardstick competition among local governments is likely to push the production and distribution of public services towards a higher level of efficiency. Nevertheless, the success of a federal setting, as far as efficiency is concerned, depends on its ability to tackle certain problems, such as the spillover effects, the risk of cost shifting and of unexploited economies of scale. On the other hand, the overall equity of the system needs to be maintained through effective policy instruments of redistribution.

It is not straightforward to assess the impact of decentralization on the efficiency and equity of the Swiss health system due to a lack of counterfactual evidence, since the health system has come with such an organization of government ever since its origins. Nevertheless, by relying on the literature, we were able to identify some of the main challenges that Switzerland will have to face in the future.

Firstly, the supply of health services is not always targeted at the most efficient dimension; this is true for hospitals and nursing homes, for example. Consortia of cantons or of municipalities may be advised to better exploit economies of scale. In 2009, a new attempt to concentrate the highly specialized medicine (that is, complex surgical procedures with rather low frequency, such as stroke units and pediatric oncology) started under the leadership of the Conference of Cantonal Health Ministers. A public call for a given set of procedures was launched; hospitals prepare and submit dossiers (in which they must disclose the number of interventions carried out in the previous year and the related rate of success), while the final decision is made based on both a technical assessment and a political appraisal. Secondly, the level of health expenditures varies enormously across cantons and a higher expenditure does not seem to systematically reflect a higher quality of service. Cantons with higher expenditure should choose policies that aim to reduce costs (for example, by improving the acceptance and availability of managed care plans) and should make greater efforts to keep the medical supply under control in order to better counter the phenomenon of supply-induced demand.

Finally, equity in financing is far from having been reached; the autonomy given to cantons to choose their subsidy policy led to a huge level of heterogeneity among them. This shows a concrete risk of "race to the bottom" that may undermine the overall equity of the system. There is still a need for research at the cantonal level for the other equity aspects, such as the equity of health and equity of the access to healthcare, although the average indicators computed for Switzerland as a whole do not assess these areas as some of main concern.

The real novelty in Switzerland seems to be the new activism of the Federal State in the health policy arena. The two OECD/WHO reports of 2006 and 2011 argued that the highly fragmented governance and the lack of leadership were the major challenges of the Swiss health system. Smith et al. (2012) articulated the concept of leadership in the three essential tasks of setting priorities, performance measurement, and designing accountability mechanisms. In 2013, the minister of home affairs launched, for the first time in Switzerland's history, a health policy agenda (called "Health2020") that sets priorities for health-policy action in four areas and defines 36 measures to be implemented in the coming 8 years.[16] Among the addressed problems are, not surprisingly, the objectives of "reinforcing fair funding and access" and "increasing efficiency" (by 20 %), particularly in the cantons with the lowest financing equity or the highest healthcare spending; these are both issues that are at the heart of the present chapter. The reactions of stakeholders and cantonal authorities to this initiative were mixed and it is too early to assess the likelihood of it being accepted and successfully implemented.[17] However, it demonstrates (as was observed in the US with the Affordable Care Act) that even federal states with a longstanding tradition of decentralization need, at a certain point in their history, to overcome fragmentation, a certain lack of governance, and weak health policy leadership, and rely increasingly on central power interventions. Of course this central power is not meant as rigid, top-down, monocentric hierarchy, but as one key-actor of the polycentric approach to governance discussed by Elinor Ostrom (2005), which was implemented in Switzerland in the context of healthcare.

References

Achtermann, W., & Berset, C. (2006). *Gesundheitspolitiken in Der Schweiz-Potential Für Eine Nationale Gesundheitspolitik. Analysen Und Perspektiven* (Bd. 1). Bern: Bundesamt für Gesundheit.

Alber, J., & Bernardi-Schenkluhn, B. (1992). *Westeuropäische Gesundheitssysteme Im Vergleich: Bundesrepublik Deutschland.* Schweiz, Frankreich, Italien, Großbritannien: Campus.

Allin, S., Masseria, C., & Mossialos, E. (2009). Measuring socioeconomic differences in use of health care services by wealth versus by income. *American Journal of Public Health, 99*(10), 1849–1855.

Banting, K. G., & Corbett, S. M. (2002). *Health policy and federalism: A comparative perspective on multi-level governance*. Kingston: Queen's University, Institute of Intergovernmental Relations.

Bilger, M. (2008). Progressivity, horizontal inequality and reranking caused by Health System Financing: A decomposition analysis for Switzerland. *Journal of Health Economics, 27*(6), 1582–1593.

Bordignon, M., & Turati, G. (2009). Bailing out expectations and public health expenditure. *Journal of Health Economics, 28*(2), 305–321.

Brown, C. C., & Oates, W. E. (1987). Assistance to the poor in a federal system. *Journal of Public Economics, 32*(3), 307–330.

Brunner, H. H., Cueni, S., & Januth, R. (2007). Krankenversicherung. In G. Kocher & W. Oggier (Eds.), *Gesundheitswesen Schweiz 2007–2009. Eine Aktuelle Übersicht* (pp. 151–170). Bern: Verlag Hans Huber.

Cheng, T. M. (2010). Understanding the 'Swiss watch'function of Switzerland's Health System. *Health Affairs, 29*(8), 1442–1451.

---

[16] See http://www.bag.admin.ch/gesundheit2020/index.html?lang=en.

[17] In fact, this activism of the federal government still lacks a constitutional basis.

Costa-Font, J. (2012). Fiscal federalism and European health system decentralization: A perspective. LEQS paper No. 55/2012.
Crivelli, L. (1998). Cross-border care between Swiss cantons: A testing lab for the Single European Market. In R. Leidl (Ed.), *Healthcare andits Financing in the Single European Market*. Amsterdam: Kluwer.
Crivelli, L., & Filippini, M. (2003). Il Federalismo Nel Settore Sanitario. In A. Ghiringhelli (Ed.), *Il Ticino Nella Svizzera* (pp. 353–379). Locarno: Dadò.
Crivelli, L. (2014). Consumer-driven health insurance in Switzerland, where politics is governed by federalism and direct Democracy. In S. Thomson & E. Mossialos (Ed.). *In private health insurance and medical savings accounts: History, politics, performance*, Forthcoming ed., Cambridge: Cambridge University Press.
Crivelli, L., & Bolgiani, I. (2009). Consumer-Driven Versus Regulated Health Insurance in Switzerland. In K. Okma & L. Crivelli (Eds.), *Six countries, six reform models: The healthcare reform experience of Israel, the Netherlands, New Zealand, Singapore, Switzerland and Taiwan* (pp. 137–170). Singapore: World Scientific Publishers.
Crivelli, L., Filippini, M., & Lunati, D. (2001). Dimensione Ottima Degli Ospedali in Uno Stato Federale *Economia Pubblica. XXXI, (5)*, 97–119.
Crivelli, L., Filippini, M., & Lunati, D. (2002). Regulation, ownership and efficiency in the Swiss Nursing Home Industry. *International Journal of Health Care Finance and Economics, 2*(2), 79–97.
Crivelli, L., Filippini, M., & Mosca, I. (2006). Federalism and regional health care expenditures: An empirical analysis for the Swiss cantons. *Health Economics, 15*(5), 535–541.
Crivelli, L., & Salari, P. (2012). Fiscal federalism and income redistribution through healthcare financing: An empirical analysis for the Swiss cantons CEPRA working paper, USI Università della Svizzera italiana.
Eichenberger, R., & Frey, B. S. (2006). Functional, overlapping and competing jurisdictions (FOCJ): A complement and alternative to today's federalism. In E. Ahmed & G. Brosio (Eds.), *Handbook of Fiscal Federalism* (pp. 154–181) Cheltenham: Edward Elgar.
Enthoven, A. C. (1993). The history and principles of managed competition. *Health Affairs, 12*(suppl 1), 24–48.
Farsi, M. & Filippini, M. (2006). An analysis of efficiency and productivity in Swiss hospitals. *Schweizerische Zeitschrift für Volkswirtschaft und Statistik, 142*(1), 1–37.
Federal Office of Public Health. (2013). *Health 2020*. Bern: Federal Office of Public Health.
Feld, L. P. (2000). Tax competition and income redistribution: An empirical analysis for Switzerland. *Public Choice, 105*(1–2), 125–164.
Gerritzen, B. C., & Kirchgässner, G. (2013). Federalism in health and social care in Switzerland. In J. Costa-Font & S. L. Greed (Eds.), *Federalism and decentralization in European Health and Social Care* (pp. 250–271). Basingstoke: Palgrave Macmillan.
Gilardi, F., & Füglister, K. (2008). Empirical modeling of policy diffusion in federal states: The Dyadic approach. *Swiss Political Science Review, 14*(3), 413–450.
Gilliand, P. (1990). *Assurance-Maladie – Quelle Révision?*. Lausanne: Éditions Réalités Sociales ed.
Gilliand, P. (1986). *Les Coûts Et l'assurance*. Lausanne: Éditions Réalités Sociales ed.
Kifmann, M., & Roeder, K. (2011). Premium subsidies and social health insurance: Substitutes or complements? *Journal of Health Economics, 30*(6), 1207–1218.
Kocher, G., & Oggier, W. (2007). *Gesundheitswesen Schweiz 2007–2009*. Bern: Verlag Hans Huber.
Leu, R. E., Rutten, F., Brouwer, W., Matter, P., & Rütschi, C. (2007). *A tale of two systems: The Swiss and the Dutch health care systems compared*. Baden-Baden: Nomos Verlagsgesellschaft.
Leu, R. E., & Schellhorn, M. (2004a). The evolution of income-related health inequalities in Switzerland over time. IZA DP No. 1346.
Leu, R. E., & Schellhorn, M. (2004b). The evolution of income-related inequalities in health care utilization in Switzerland over time. IZA Discussion paper series 1316.

Levaggi, R., & Zanola, R. (2004). Patients' migration across regions: The case of Italy. *Applied Economics, 36*(16), 1751–1757.
Musgrave, R. A. (1971). Economics of fiscal federalism. *Nebraska Journal of Economics and Business*, 3–13.
Nolte, E., & McKee, M. (2011). Variations in amenable mortality—trends in 16 high-income Nations. *Health Policy, 103*(1), 47–52.
O'Donnell, O., Van Doorslaer, E., Wagstaff, A., & Lindelow, M. (2008). *Analyzing health equity using household survey data*. Washington DC: The World Bank Publication.
Oates, W. E. (1972). *Fiscal federalism*. New York: Harcourt Brace Jovanovich ed.
Oates, W. E. (1999). An essay on Fiscal federalism. *Journal of Economic Literature, 37*(3), 1120–1149.
Oates, W. E. (2005). Toward a second-generation theory of Fiscal federalism. *International Tax and Public Finance, 12*(4), 349–373.
OECD/WHO. (2011). *OECD reviews of health systems: Switzerland 2011*. OECD publishing.
OECD/WHO. (2006). *OECD reviews of health systems: Switzerland 2006*. OECD publishing.
Ostrom, E. (2005). Unlocking public entrepreneurship and public economies UNU-WIDER Discussion paper No. 2005/01.
Pauly, M. V. (1973). Income redistribution as a local public good. *Journal of Public Economics, 2*(1), 35–58.
Reich, O., Weins, C., Schusterschitz, C., & Thöni, M. (2012). Exploring the disparities of regional health care expenditures in Switzerland: Some Empirical evidence. *The European Journal of Health Economics, 13*(2), 193–202.
Reinhardt, J. D., von Elm, E., Fekete, C., & Siegrist, J. (2012). Social inequalities of functioning and perceived health in Switzerland—A representative cross-sectional analysis. *PLoS ONE, 7*(6), e38782.
Schaffhauser, R., Locher, H., Poledna, T., & Finger, M. P. (2006). Das Gesundheitswesen: Motor Von Wohlbefinden Und Wohlstand: Radikale Denkanstösse Für Das Schweizerische Gesundheitswesen: Institut für Rechtswiss. und Rechtspraxis.
Smith, P. C., Anell, A., Busse, R., Crivelli, L., Healy, J., Lindahl, A. K., Westert, G., & Kene, T. (2012). Leadership and governance in seven developed health systems. *Health Policy, 106*(1), 37–49.
Stiglitz, J. E. (1988). *Economics of the public sector*. New York: WW Norton.
Tiebout, C. M. (1956). A pure theory of local expenditures. *Journal of Political Economy, 64*(5), 416–424.
Van Doorslaer, E., Wagstaff, A., & Rutten, F. (1992). *Equity in the finance and delivery of health care: An international perspective*. Oxford: Oxford University Press.
Van Doorslaer, E., Wagstaff, A., Van der Burg, H., Christiansen, T., De Graeve, D., Duchesne, I., et al. (2000). Equity in the delivery of health care in Europe and the US. *Journal of Health Economics, 19*(5), 553–583.
Vatter, A., & Rüefli, C. (2003). Do political factors matter for health care expenditure? A comparative study of Swiss cantons. *Journal of Public Policy, 23*(3), 301–323.
Wagstaff, A., Van Doorslaer, E., Van der Burg, H., Calonge, S., Christiansen, T., Citoni, G., et al. (1999). Equity in the finance of health care: Some further international comparisons. *Journal of Health Economics, 18*(3), 263–290.
Widmer, P., & Zweifel, P. (2012). Fiscal equalization, Tiebout competition, and incentives for efficiency in a federalist country. *Public Finance Review, 40*(1), 3–29.
World Health Organization. (2000). *The World Health Report 2000: Health Systems: Improving performance*. WHO.
Wyss, K., & Lorenz, N. (2000). Decentralization and central and regional coordination of health services: The case of Switzerland. *The International Journal of Health Planning and Management, 15*(2), 103–114.
Zweifel, P., Telser, H., & Vaterlaus, S. (2006). Consumer resistance against regulation: The case of health care. *Journal of Regulatory Economics, 29*(3), 319–332.

Patients' Mobility Across Borders: A Welfare Analysis

Laura Levaggi and Rosella Levaggi

Abstract Welfare systems are designed on geographical and membership boundaries. In terms of access to health care this implies that, as a general rule, only individuals residing in their national territory can obtain health care from providers located there. However, in the past few years medical tourism has grown at an explosive pace throughout the world and in Europe. Each year in fact a small, but significant number of European citizens seek medical treatment that is financed by their public insurer in another EU country. From an economic point of view, it is important to distinguish between the two following sources of patients' mobility: a regulated mobility, where the third payer decides to send patients abroad and patients' choice, where the patient himself decides to seek care abroad. In this article we show how the combined effect of restrictions to the use of health care, transfer prices, and mobility rules determine social welfare and its allocation between Regions. The results are quite interesting: if the price set for these patients is equal to the marginal cost of the more efficient Region, patients' mobility should be preferred to patients' choice. On the other hand, if the price is equal to the marginal cost of the less efficient Region, patient choice should be preferred. The other interesting result is a possible trade off between a static model where each Region chooses its level of cost/effectiveness and a more long term situation, where patient mobility determines a common level for this parameter.

Keywords Health care provision · Patients' mobility · Short run equilibrium · Long run equilibrium

L. Levaggi
Faculty of Science and Technology, Free University of Bolzano/Bozen,
Piazza Università 1, 39100 Bolzano, Italy
e-mail: levaggi@unibz.it

R. Levaggi
Department of Economics and Management, University of Brescia,
Via San Faustino 74/b 25122 Brescia, Italy

R. Levaggi and M. Montefiori (eds.), *Health Care Provision and Patient Mobility*, 179
Developments in Health Economics and Public Policy 12,
DOI: 10.1007/978-88-470-5480-6_8, © Springer-Verlag Italia 2014

JEL Classification I18 · I11 · H42 · H51

1 Introduction

Since their inception, welfare systems have been constructed on geographical and membership boundaries. In terms of access to health care this implies that, as a general rule, only individuals residing within the national boundary can obtain health care from providers located there. In this way member states can control quality of care, protect the financial stability of the national system, and ensure adequate planning of health care infrastructure and capacity. However, in the past few years medical tourism has grown at an explosive pace throughout the world. The driving force behind this phenomenon is certainly represented by American insurances that offer reductions to policy-holders who are willing to be treated in other countries. The US health care system is in fact one of the most expensive worldwide and significant savings can be obtained by being treated in low to middle income countries, but also in Europe. Each year an increasing number of patients[1] travel to other countries to affort health care treatments (especially ortopaedics and cardiac) and their most preferred destination is Asian and Central American countries where medical centers devoted to treat these patients are set up. In publicly driven health care systems, cross border mobility is an even more recent phenomenon that can take several forms. Acute and emergency health care as a consequence of mobility throughout Europe, is regulated by specific rules, but it is a very marginal phenomenon. In the past few years another and more significant flow has been emerging. Each year in fact a small, but significant number of European citizens seek medical treatment financed by their public insurer in another EU country. From an economic and a social point of view it is important to distinguish between the following two sources of patient mobility:

1. Third party insurer's decision: in this case the purchaser of health care (public or private) encourages a set number of patients to receive care abroad, either because it may be less expensive or in order to face a temporary unbalance between demand and supply. In the UK the NHS has encouraged patients to receive care in France, while Belgium, The Netherlands and Luxembourg have outsourced patients to neighbouring countries (Glinos et al. 2006; Burge et al. 2004; Legido-Quigley et al. 2007).
2. Patient choice: in this case the patient himself decides to seek care abroad. The reason for such choice may lie in the type of treatment offered, in the perceived level of quality, or in the waiting time differential. In the second case the rules about rights and reimbursement are far from clear and the European Court of Justice has often been asked to rule on disputes between citizens and member states. A new directive has been proposed and has come into force in 2011, but

[1] Estimates vary from half to two million patients. See Alleman et al. (2011) for a review.

its process is still slow. In fact member States with a negative mobility flow are opposing this Directive[2]. In spite of its importance both from an economic, social and legal point of view, the impact that patients' mobility may have on welfare states has not been studied. Brekke et al. (2011) show that patients' mobility is not necessarily welfare improving and that benefits from such mobility may not be equally distributed among countries.

In this article we show how the combined effect of restrictions to the use of health care, transfer prices, and mobility rule may determine social welfare and its allocation between Regions. In particular we will study and compare the welfare effects of regulated patient mobility (point 1 above) and patient choice (point 2). The results are interesting. If the transfer price set for these patients is equal to the marginal cost of the more efficient Region, patient mobility should be preferred to patient choice. On the other hand, if the price is equal to the marginal cost of the less efficient Region, patient choice should be preferred. The other interesting result is a possible trade off between a static model where each Region chooses its level of cost effectiveness and a more long term situation, where patient mobility determines a common level for this parameter. The paper will be organised as follows: in Sect. 2 we describe patient mobility, in Sect. 3 we present the model and the main conclusions of our analysis. In Sect. 4 long-run considerations are made and conclusions are then drawn in Sect. 5.

2 Patient Mobility

Several quite different phenomena are grouped under the general heading of patient mobility. Mainil et al. (2012) have recently tried to order the semantic differences in this literature. They suggest using transnational health care as a heading for several phenomena ranging from medical tourism to receiving A&E care while abroad. The institutional settings are in fact quite important. In the US where health care is mainly provided by the private sector, patient mobility is driven by cost saving arguments. In order to reduce their insurance premiums, individuals accept to be treated outside the US, where the labor cost is much lower. In this way, health care costs can be significanlty reduced without a great loss in the quality of care. The same reasoning may apply to patients that travel within Europe for treatments that have to be paid for by the individual out of his/her pocket. In this case costs are reduced, but quality is somehow more difficult to compare. In this article we concentrate on another flow of patients, i.e. we model the decision of an individual to receive care outside his/her Region of residence, while still financed by his/her National Health care system. In Europe the number of patients involved in this decision is fairly limited and it accounts for about 1 %

[2] For a detailed account of the EU current legislation see Greer et al. (2013), Santuari (2013).

of total health expenditure Glinos et al. (2010), but it may grow significantly in the near future, especially if some of the legal and economic barriers to mobility are removed. A recent survey on cross border mobility in Germany Wagner and Linder (2010) shows that patients go abroad to receive care sometimes as part of (or in combination with) a holiday. The preferred destination for German patients is in fact Spain and the most frequently required treatment is related to chronic back or join pain. At the moment, only health care expenditure can be reimbursed while traveling expenses for the patient and his/her family have to be borne by the patient.

Patient mobility involves legal and economic aspects. From a legal point of view, it is necessary to define whether patient mobility falls under free movement of people or free movement of goods. This is an important distinction because it delimits the power member States have in restricting patient choice.

In the former case, the patient can decide where to be treated and his/her health care system has to reimburse expenses, in the latter case it is possible to curb such mobility by asking for some forms of permission to receive care abroad. The European Court of Justice has been asked to rule on several controversies between citizens and their Member States concerning the reimbursement of health care. From these rulings some principles seem to emerge Legido-Quigley et al. (2007):

- All EU citizens can claim outpatient treatment to which they would be entitled in their own country in any other EU member state without prior permission.
- All EU citizens can claim inpatient treatment to which they would be entitled in their own country in any other EU member state subject to prior permission. The authorization can be refused on the ground of lack of medical necessity only if the same or equally effective treatment can be obtained without undue delay in the country of residence.

In both cases, the costs must generally be reimbursed up to the amount which would be reimbursed in the patient's home country if the domestic preconditions are fulfilled. Several aspects should be studied and in this analysis we concentrate on the welfare properties of the patient flow. In particular, we will compare the decision of the regulator with the decision of the patient to receive health care abroad. To do so we will use a very simple model that combines spatial competition with health care expenditure decisions.

3 The Model

In our model we define regulated mobility (RM) as the flow of patients from one Region to another that originates from a Regulator's decision. In this case patients are asked to go outside their Region to receive care. The mobility costs are borne by the regulator and the patient is often incentivated to make this choice either because this is the only alternative (that specific service is not offered in their

Region of residence) or by some quality related issues (for example a shorter waiting list). Patient choice (PC), on the other hand, is the flow of patients that originates from the patients' decision to seek a provider outside their area. Depending on each country's regulation in this case the cost for the treatment may be reimbursed, while traveling costs are not necessarily repaid. In our model we take account of this asymmetric treatment by assuming that patient utility is linear in income and health related utility. With this functional form distributional issues are not relevant and we can simply model the two decisions. In actual fact distributional issues are important and they will be discussed in the final part of the paper.

3.1 The Environment

Our community consists of a mass of individuals, normalized to one, uniformly distributed on a unit line, with a fixed exogenous income Y, distributed according to a density function f in the support $\left(\underline{M}, \overline{M}\right)$. We denote by $Y_M = \int_{\underline{M}}^{\overline{M}} Yf(Y)\,\mathrm{d}Y$ its average. The community is split into two equal Regions; in other words people located in the area $\left[0, \frac{1}{2}\right]$ belong to Region 0 and those located in $\left[\frac{1}{2}, 1\right]$ belong to Region 1. Individual utility depends on the level of income (as a proxy for the goods that can be purchased on the market) and health. The latter is supplied by two facilities located at the extremes of the unit line that are run by the Regulator in each Region. There is no uncertainty and the total cost can be written as:

$$C_j = v_j S_j + \theta_j \frac{q_j^2}{2} \qquad j = 0, 1 \tag{1}$$

where S_j is the supplied quantity, v_j captures technology aspects and minimum quality requirements, $\theta_j \frac{q_j^2}{2}$ is the cost incurred by provider j to increase quality to level q_j.

Each individual is endowed with a stock of health, normalized to zero for simplicity, that can be improved through a standard treatment supplied by the public health care system. The money equivalent utility of the health gain that a generic individual may enjoy is equal to αe, where α is a fixed parameter that represents the money utility of health, while e measures the degree of effectiveness of the treatment itself. We model the effectiveness e as a random variable uniformly distributed in the interval $[0, 1]$.[3] It depends on specific characteristics of the patients, but we assume that it is independent of the level of income. Access is restricted by the regulator: only individuals whose effectiveness e is above a pre-

[3] The effectiveness of the treatment can vary for several reasons, for example an active principle that reduces blood pressure might be less effective when used for treating patients with multiple diseases.

set threshold c_j are eligible to receive the service. This is supplied free of charge, but users incur travel costs at rate m to receive it. Users are indexed by their position on the line, so that x represents patients located at distance x from the origin. The utility derived from health care for a generic individual residing in Region j, located at x is given by:

$$\begin{cases} \alpha e + h + \phi q_0 - mx, & \text{if } e \geq c_j \text{ \& care is received in } 0, \\ \alpha e + h + \phi q_1 - m(1 - x) & \text{if } e \geq c_j \text{ \& care is received in } 1, \\ 0 & \text{if } e < c_j. \end{cases} \quad (2)$$

Here $\phi > 0$ is the evaluation of quality q_j offered by competitor j, so that ϕq_j is the monetary equivalent gain derived from using the service from provider j. The constant term h is sufficiently high to assure that any patient chooses to receive the service from some provider, i.e. the service is essential. Finally, in line with the most recent literature on patient competition (Brekke et al. 2012), we assume that $\phi = 1$; in this case m can be interpreted as the marginal cost of traveling relative to quality evaluation.

The quantity of health care that is demanded and has to be financed by Region j is thus equal to:

$$D_j = \int_{c_j}^{1} \frac{1}{2} \mathrm{d}e = \frac{1 - c_j}{2}. \quad (3)$$

Although total demand $D_0 + D_1$ must be equal to the total supply $S_0 + S_1$, the quantity demanded by residents in each local authority (D_j) does not necessarily need to coincide with the quantity produced in the same area (S_j). In other words we allow for mobility, i.e. some patients may be treated outside their Region,[4] upon the payment of a price $p \in [v_0, v_1]$.

The provision of health care is financed using a linear income tax at rate τ_j. The welfare function in each Region j can thus be written as:

$$W_j = (1 - \tau_j)Y_j + \alpha \int_{c_j}^{1} e\,\mathrm{d}e + h + \int_{\frac{1}{2}j}^{r_j} (q_0 - mx)\,\mathrm{d}x + \int_{r_j}^{\frac{1}{2}(j+1)} (q_1 - m(1 - x))\,\mathrm{d}x$$
$$\tau_j = \frac{v_j S_j + p(D_j - S_j) + \theta_j \frac{q_j^2}{2}}{Y_j}, \quad (4)$$

with Y_j being the average revenue in Region j and r_j the position of the nearest (to the inland access point) patient in Region j that will move outside it to receive

[4] The quantity produced rather than the one consumed produces utility for several reasons well explained in Wildasin (2001, 2004). In this note we simply mention the option good characteristic of health care.

care. In this context, we study patient flows between the two Regions, in particular as regards their welfare implications. As suggested in Sect. 2, these flows have different origins and should be studied separately. In our analysis we will concentrate on what we have called regulated mobility and patient choice. In the first case the provider decides how many patients to treat outside and it will take this decision using a cost-effectiveness argument, i.e. it will take into account the difference in quality and the difference in the total cost of health care by considering traveling expenses and health care costs. When, on the other hand, patients can freely choose where to go, a true competition model develops between the two Regions, but in this context more competition does not necessarily imply welfare improvement. Patients do not consider health care costs in their decision, they simply evaluate the difference between quality and traveling costs. For providers quality becomes a constraint rather than a decision variable.

It is interesting to note that the private traveling costs are borne by the patients in our model. Some health care systems discriminate between patient mobility and patient choice. When patients are sent outside their region of residence, the health care system bears mobility costs; when the decision is taken by the patient he/she usually has to pay these expenses. In our model, given that utility is linear in income, this aspect is irrelevant for welfare considerations.

3.2 Equilibrium Without Mobility

If mobility is not allowed, each Region maximizes welfare by choosing the level of quality and the number of treatments that should be provided. From (4) the problem is then to maximize over q_j the following function:

$$\begin{aligned} Y_j - \frac{v_j}{2}(1 - c_j) - \theta_j\frac{q_j^2}{2} + \frac{1}{2}\alpha\left(1 - c_j^2\right) + h(1 - c_j) \\ + (1 - c_j)\left(\frac{1}{2}q_j - \frac{1}{8}m\right) \end{aligned} \tag{5}$$

and the optimal solution is to set

$$q_{j,NM} = \frac{1 - c_j}{2\theta_j}. \tag{6}$$

Below we will assume that the optimal number of individuals that are allowed to receive health care is set using a cost effectiveness criterion. Substituting (6) in (5) the optimal welfare level is thus

$$\begin{aligned} W_{j,NM} = {} & Y_j - \frac{1}{2}v_j(1 - c_j) - \frac{1}{8}\frac{(1 - c_j)^2}{\theta_j} \\ & + \frac{1}{2}\alpha\left(1 - c_j^2\right) + h(1 - c_j) + \frac{1 - c_j}{4}\left(\frac{1 - c_j}{\theta_j} - \frac{m}{2}\right). \end{aligned} \tag{7}$$

3.3 *Patient Mobility*

In this subsection we assume that mobility is allowed. In this first subsection we assume that the regulator can control patient mobility, i.e. patients are required to receive care abroad, while in the second, patients can decide their preferred provider.

3.3.1 Regulated Mobility

Below we assume that each Region can control the flow of patients that go abroad. There are several reasons that induce decision-makers to take this decision: it may be a problem related to underprovision, excess demand or it may simply be that the neighbouring Region is more efficient at producing health care. In our analysis we will focus on this latter aspect, i.e. we assume that patient flows are determined by an imbalance in the productivity level of health care at two levels: the current cost and the cost to produce quality. In order to achieve a better understanding of the results, we assume that Region 0 is more productive in both measures, i.e.

$$v_0 < v_1, \text{ and } \theta_0 < \theta_1. \tag{8}$$

The demand for health care in each local authority still depends on c_j, but local supply depends on the number of patients that move. In our setting this will be defined by the position of the moving patients on the line. Let us define the borderline position by r. By hypothesis (8) $r > \frac{1}{2}$ and patients will only move from Region 1 to Region 0. Local supply can then be written as:

$$\begin{aligned} S_0 &= \frac{1}{2}(1 - c_0) + (1 - c_1)\left(r - \frac{1}{2}\right) \\ S_1 &= (1 - c_1)(1 - r). \end{aligned} \tag{9}$$

Let us analyze the behavior of each local authority. In this case the decision of Region 1 does not affect the choices of Region 0. The problem can then be solved either as a simultaneous or a sequential equilibrium.

3.3.2 Region 0

Region 0 decides how many patients receive from Region 1 and its level of quality q_0. The number of its own patients to be treated depends on c_0 and is supposed to be already fixed; we will go back to define its level in the policy implications of the model. The objective of Region 0 is thus to maximize

$$Y_0 - \frac{1}{2}v_0(1-c_0) - (v_0-p)(1-c_1)\left(r-\frac{1}{2}\right) - \frac{\theta_0 q_0^2}{2} + \frac{\alpha(1-c_0^2)}{2} + h(1-c_0) + (1-c_0)\int_0^{\frac{1}{2}}(q_0 - mx)\mathrm{d}x. \tag{10}$$

Since $p \geq v_0$ the function is strictly increasing in r: given that the marginal cost to treat patients is constant, the Region is open to receive as many patients as Region 1 will send. For the quality, the optimal solution is then the same as before

$$q_{0,RM} = q_{0,NM} = \frac{1-c_0}{2\theta_0}. \tag{11}$$

Opening to patients from the neighbouring Region does not affect the choice of quality for the more productive Region, at least in a context where the flow is not controlled by patients themselves.

3.3.3 Region 1

The objective of Region 1, on the other hand, is to maximize over r and q_1 the following utility function:

$$Y_1 - v_1(1-r)(1-c_1) - p(1-c_1)\left(r-\frac{1}{2}\right) - \frac{1}{2}\theta_1 q_1^2 + \frac{1}{2}\alpha(1-c_1^2) + h(1-c_1) + (1-c_1)\left(\int_{\frac{1}{2}}^{r}(q_0 - mx)\mathrm{d}x + \int_r^1(q_1 - m(1-x))\mathrm{d}x\right). \tag{12}$$

The optimization problem admits an internal solution for any $p \in [v_0, v_1]$ only if

$$m > \max\left(\frac{1-c_1}{2\theta_1}, \frac{1-c_0}{2\theta_0} + v_1 - v_0\right) \tag{13}$$

and this condition will be assumed to hold throughout the rest of the paper. Under this condition the optimal values are given by:

$$q_{1,RM} = \frac{(1-c_1)(m - q_{0NM} + p - v_1)}{2m\theta_1 + c_1 - 1} \tag{14}$$

$$r_{RM} = \frac{1}{2} + \frac{v_1 - p}{2m} + \frac{q_{0,RM} - q_{1,RM}}{2m} = 1 - \frac{(m - q_{0NM} + p - v_1)\theta_1}{2m\theta_1 + c_1 - 1}. \tag{15}$$

As shown above, the optimal number of patients that will receive health care outside their own region is determined by the two aspects we anticipated in Sect. 3.1. The first term is a comparison of health care savings and traveling costs, the second one is an evaluation of the utility differential versus the cost to receive care outside.

Since $q_{1,RM} = \frac{1-c_1}{\theta_1}(1 - r_{RM})$ we also have

$$q_{1,NM} - q_{1,RM} = \frac{1 - c_1}{\theta_1}\left(r_{RM} - \frac{1}{2}\right) > 0$$

which means that the offered quality in Region 1 will be decreased in this case, compared to the situation where mobility is not allowed.

3.3.4 Welfare Analysis

The welfare can be found by substituting back into Eqs. (10) and (12) the optimal values (11), (14) and (15). Let us now examine then the welfare differential between the "No Mobility" case (NM) and the present "Regulated Mobility" (RM):

$$W_{0,RM} - W_{0,NM} = (p - v_0)(1 - c_1)\left(r_{RM} - \frac{1}{2}\right) \tag{16}$$

$$\begin{aligned} W_{1,RM} - W_{1,NM} = {} & (v_1 - p)(1 - c_1)\left(r_{RM} - \frac{1}{2}\right) - \frac{\theta_1}{2}\left(q_{1,RM}^2 - q_{1,NM}^2\right) \\ & + (1 - c_1)\left(q_{0,NM} - q_{1,NM}\right)\left(r_{RM} - \frac{1}{2}\right) \\ & + \left(q_{1,RM} - q_{1,NM}\right)(1 - r_{RM}) \\ & - \frac{m}{4}\left(6r_{RM}^2 - 8r_{RM} + 3\right). \end{aligned} \tag{17}$$

The welfare difference for Region 0 is obviously non-negative, since $p \geq v_0$. As for Region 1, since the quality offered by Region 0 does not change between the two cases, the welfare has to increase, otherwise Region 1 would set $r = \frac{1}{2}$ and get the same welfare level as in the first case. Note, however, that not all terms in (17) are positive, because as seen above the quality in Region 1 decreases, thus fixed costs are lowered, but of course this will also diminish the welfare component relative to the quality of care. In the same way, patients accessing care in 0 will get a higher quality, but will bear higher transportation costs.

3.4 Patient Choice

In this section we assume that patients may decide their preferred provider on the basis of the quality, net of the transport cost. The two Regions will then compete on quality to get patients. It is interesting to note that although in both cases mobility is allowed, the decision is rather different. In the first case the regulator takes into account both the quality gain, and the lower cost of the treatment. In this second case, given that we have assumed that in the public health care system care is free at the point of use, patients will choose on the basis of quality and private costs. For this reason, the position of the patient that is indifferent between provision in his/her Region and going outside will be represented by

$$z = \frac{1}{2} + \frac{1}{2m}(q_0 - q_1). \tag{18}$$

In the comparison with Eq. (15), representing the optimal number of patients that move when the Government can decide, we can immediately recognize this difference in the term $\frac{v_1 - p}{2m}$. In this model the patient's decision becomes a constraint for the decision makers which they need to take into account in order to maximize total welfare. Let us now examine this case in detail.

3.4.1 Region 0

Region 0, by setting its preferred level of quality q_0 can compete with Region 1 for patients and the number of moving people will depend on the difference between the two qualities. The optimal choice will be determined by maximizing the welfare function:

$$Y_0 - v_0\frac{1-c_0}{2} + (p - v_0)(1-c_1)\frac{1}{2m}(q_0 - q_1) - \frac{\theta_0 q_0^2}{2} + \frac{\alpha}{2}(1 - c_0^2) + h(1-c_0) + (1-c_0)\int_0^{\frac{1}{2}}(q_0 - mx)\mathrm{d}x. \tag{19}$$

The optimal choice for q_0 is equal to

$$q_{0,PC} = \frac{1-c_0}{2\theta_0} + \frac{(p-v_0)(1-c_1)}{2\theta_0 m} = q_{0,NM} + \frac{1}{2\theta_0}\frac{(p-v_0)(1-c_1)}{m}. \tag{20}$$

The second term in (20) is non-negative, hence we can conclude that in equilibrium the quality produced by Region 0 is going to be higher than in the previous cases, unless $p = v_0$. This is an important difference compared to the previous model and it shows the different effect that patient choice has in this case. There the extra payment made to the Region was used to increase income, here by attracting patients from outside it reduces production costs and the benefit is passed on to its own patients in terms of a higher quality.

3.4.2 Region 1

Region 1, on the other hand, will maximize over q_1 the following utility function:

$$Y_1 - v_1(1-c_1)(1-z) - p(1-c_1)\left(z-\frac{1}{2}\right) - \frac{1}{2}\theta_1 q_1^2 + \frac{1}{2}\alpha(1-c_1^2) + h(1-c_1) + (1-c_1)\left(\int_{\frac{1}{2}}^{z}(q_0 - mx)\mathrm{d}x + \int_{z}^{1}(q_1 - m(1-x))\mathrm{d}x\right) \tag{21}$$

with z as in (18). The optimal value is:

$$q_{1,PC} = \frac{(1-c_1)\left(m - v_1 + p - q_{0,PC}\right)}{2m\theta_1 + c_1 - 1}. \tag{22}$$

Note that the dependence on the quality in Region 0 is the same as in (14). Since by (13) $2m\theta_1 + c_1 - 1 > 0$ and $q_{0,PC} \geq q_{0,RM}$ we have $q_{1,PC} \leq q_{1,RM}$ and the quality in Region 1 is further reduced whenever $p > v_0$.

3.4.3 Equilibrium

In equilibrium the number of moving patients will then be

$$z_{PC} = \frac{1}{2} + q_{0,NM}\frac{\theta_1}{-1 + c_1 + 2m\theta_1} + \frac{1}{2}\frac{(1-c_1)(\theta_1(p - v_0) + \theta_0(v_1 - p - m))}{(-1 + c_1 + 2m\theta_1)\theta_0 m}. \tag{23}$$

Comparing (23) with (15) we see that for $p = v_0$ the difference $z_{PC} - r_{RM}$ is negative and the number of patients that will move in PC will be less than in RM, while for $p = v_1$ is positive and the situation is reversed. In fact it can be shown that $z_{PC} - r_{RM}$ is positive whenever p is greater than the following convex combination of v_0 and v_1:

$$\frac{\theta_1(1-c_1)}{\theta_0(2m\theta_1 - 1 + c_1) + \theta_1(1-c_1)}v_0 + \frac{\theta_0(2m\theta_1 - 1 + c_1)}{\theta_0(2m\theta_1 - 1 + c_1) + \theta_1(1-c_1)}v_1. \tag{24}$$

Note also that for $p = v_0$ the quality offered by both Regions is the same as in the previous case, but the number of patients that move is lower than the optimal one and welfare will be reduced in PC as compared to RM.

3.4.4 Welfare Analysis

The optimal welfare level is found by substituting back into Eqs. (19) and (21) the optimal quantities (20), (22) and (23). As in the previous case, we examine the welfare difference between "No Mobility" (NM), "Regulated Mobility" (RM) and "Patients Choice" (PC). In contrast with the previous section, the comparison here is not so straightforward, since the optimal values in the two cases result from different maximization processes.

Let us first examine the welfare difference for Region 0:

$$W_{0,PC} - W_{0,RM} = (p - v_0)(1 - c_1)(z_{PC} - r_{RM}) - \frac{\theta_0}{2}\left(q_{0,PC}^2 - q_{0,NM}^2\right) + \frac{(1 - c_0)}{2}\left(q_{0,PC} - q_{0,NM}\right). \quad (25)$$

As already noted above, the sign of the first term depends on the price p, the second is always negative, since quality is raised, while the third is positive. For the extreme cases $p = v_0$ and $p = v_1$ it is easy to show that the difference is zero in the first case and positive in the second. The relationship with p is not monotone; it can in fact be shown that the difference in (25) is positive whenever p is greater than the following convex combination of v_0 and v_1:

$$\frac{(1 - c_1)(2\theta_1 m + 1 - c_1)v_0 + 4m\theta_0(2m\theta_1 - 1 + c_1)v_1}{8m^2\theta_1\theta_0 - 4m(1 - c_1)\left(\theta_0 - \frac{1}{2}\theta_1\right) + (1 - c_1)^2}. \quad (26)$$

The above threshold can be compared with (24): since

$$\frac{1 - c_1 + 2m\theta_1}{8m^2\theta_1\theta_0 - 4m(1 - c_1)\left(\theta_0 - \frac{1}{2}\theta_1\right) + (1 - c_1)^2} < \frac{\theta_1}{\theta_0(2m\theta_1 - 1 + c_1) + (1 - c_1)\theta_1}$$

the price threshold necessary to increase welfare in Region 0 from RM to PC is greater than the one to increase the number of treated patients coming from Region 1, i.e. in PC the same level of welfare is reached only by treating more patients.

The comparison between the welfare levels for Region 0 in PC and NM, is much simpler. Since for $p > v_0$ we have $q_{0,PC} > q_{0,NM}$ and $W_{0,PC}$ will always be greater than $W_{0,NM}$. In fact by the hypotheses Region 0 will never lose its own patients, thus the part of the welfare function deriving from access to health care will not decrease and the same holds for the first one, otherwise it would make no sense to increase the quality to get patients from Region 1.

Let us now turn to Region 1 and examine first the welfare in PC and RM. In the first case patients move on the basis of a comparison between the welfare gain they get from enhanced quality and transportation costs, while in the second Region 1 decides how many patients to send outside by also considering costs. The lower the price p, the more expedient it will be for Region 1 to send patients outside, but at the same time the lower the quality that will be offered in PC by Region 0, thus the lower the number of patients that will really move in PC. And in fact for the extreme case $p = v_0$ we have $q_{0,PC} = q_{0,RM} = q_{0,NM}$, while the number of patients that move from 0 to 1 is less than the optimal one since $z_{PC} - r_{RM} = \frac{v_0 - v_1}{2m} < 0$, thus in this case welfare will decrease. It can further be shown that $W_{1,PC} - W_{1,RM}$ is concave in p and positive for $p = v_1$. Thus for $p \in (v_0,\ v_1)$ there exist a unique value of p exists such that the two welfare functions are equal. For lower values of p, RM will give a higher welfare; for higher values, PC is the best choice. Also, by concavity an optimal value will exist for p giving the highest welfare difference. The analytic expressions of these thresholds are however rather involved.

Let us now compare the welfare levels for Region 1 in PC and NM we have

$$\begin{aligned} W_{1,PC} - W_{1,NM} &= (v_1 - p)(1 - c_1)\left(z_{PC} - \frac{1}{2}\right) - \frac{1}{2}\theta_1\left(q_{1,PC}^2 - q_{1,NM}^2\right) \\ &\quad + (1 - c_1)\left(q_{0,PC}\left(z_{PC} - \frac{1}{2}\right) - \frac{1}{2}m\left(z^2 - \frac{1}{4}\right) + q_{1,PC}(1 - z_{PC})\right. \\ &\quad \left. - \frac{1}{2}m(1 - z_{PC})^2 - \frac{1}{2}q_{1,NM} + \frac{1}{8}m\right). \end{aligned}$$

Since it can be shown that $W_{1,PC}$ is strictly decreasing in p and that $W_{1,PC} > W_{1,NM}$ for $p = v_1$, the inequality is valid for all p in the considered interval.

3.5 Numerical Example

In this section we present a numerical example in order to visualize the above results. The following parameters have been chosen for the simulation: $Y_0 = 10$, $Y_1 = 8$, $v_0 = 3$, $v_1 = 4$, $\theta_0 = 1.5$, $\theta_1 = 1.8$ and $\alpha = 10$. This implies that $q_{0,NM} = 0.2\overline{3}$ and in order to satisfy (13) m has to be chosen greater than $1.2\overline{3}$. Here we have chosen $m = 2$ and $h = m/8$. The minimum efficacy level has been selected, based on a cost/efficacy criterion, as $c_i = \frac{v_i}{\alpha}$. Substituting in (24) and (26) the number of moving patients in RM and PC is the same for $p \approx 3.9$, while the welfare in Region 0 for RM equals that for PC if $p \approx 3.94$, as shown also in Fig. 1. From the graph of the optimal welfare levels for Region 1 it emerges out that PC equals RM for $p \approx 3.761$.

This means that in general patient choice should be preferred to regulated mobility only if p is sufficiently close to the price of the inefficient region, which also corresponds to the maximum price for which mobility is allowed. This result may seem counter-intuitive: the lower the price the fewer patients should be left free to choose, but it can be easily explained. In the regulated mobility case the more efficient Region uses the extra money it receives from the inflow of patients to increase local income; when patients may choose their provider, it will have to use it also to improve quality if it wants to attract patients. In our model, as in most of this literature, quality is assumed to be a fixed cost, i.e. it does not depend on the

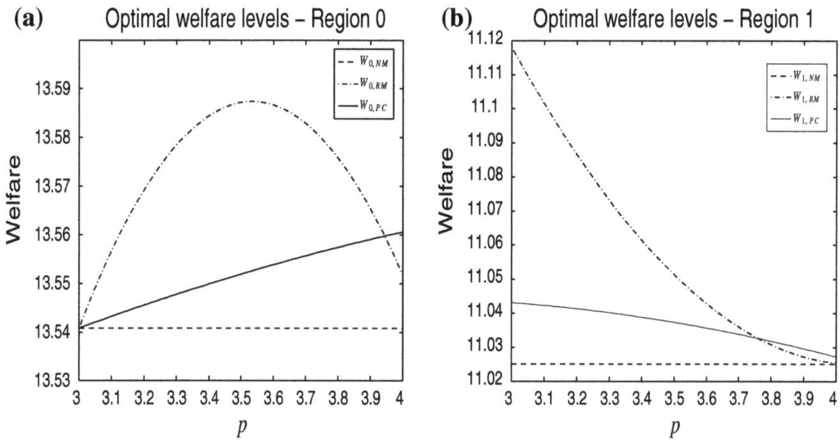

Fig. 1 Optimal welfare levels

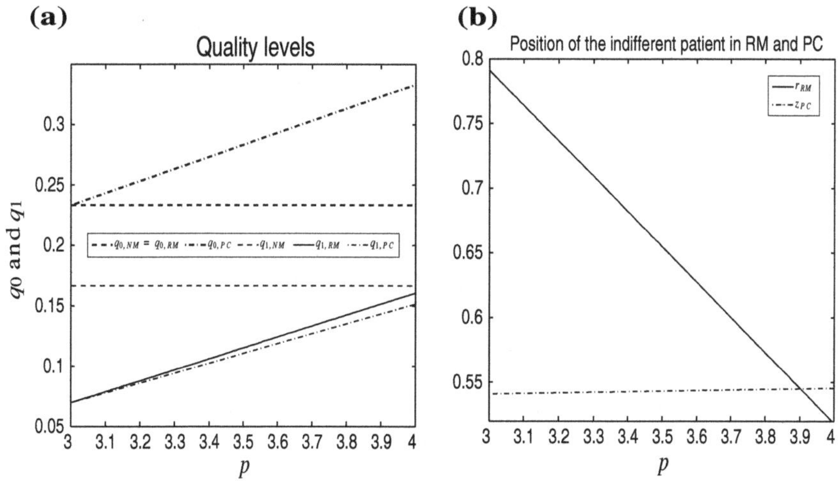

Fig. 2 Offered quality levels and number of moving patients

number of patients treated. Patients living in Region 1 will benefit from this improved quality and this will increase patient welfare. On the other hand, patient competition and a high price means that Region 1 cannot decrease local quality too much if it wants to attract patients and this increases the quality also of the patients that do not move (Fig. 2).

3.6 Discussion

The model presented in this paper highlights some important facts about patient mobility. The first interesting result is that competition is not necessarily welfare improving in this context and that gains from such mobility may not be shared equally between the two Regions.

In the previous section we have considered a short-run equilibrium where the decision of how many patients are allowed to be treated is not affected by the mobility flow; for this reason each Region sets restrictions to access based on a cost effectiveness parameter that takes into account local conditions, in particular it will consider the marginal cost to produce health care locally. The most interesting result we obtain is that while mobility always improves welfare, its form (regulated or free) depends on the price set for such flow. The numerical example proposed shows that the price should be fairly close to the cost of the less efficient Region and in this respect the EU rule to set a price equal to the cost paid in the origin Region is correct. It should however be pointed out that in our model we have considered a special case where one of the two Regions is "more efficient" both in terms of technology and cost reduction. Another interesting case to analyze would be to compare two Regions where technologies have different costs and different level of effectiveness. In this case we may have one Region (say 1) that produces health of a (relatively) low quality at a (relatively) low cost and another Region (say 0) where the cost is higher, but quality is produced more efficiently. For this case, representing for example new high-tech treatments, patient mobility in general may have different welfare properties. In our work we have not dealt with this case for two main reasons: most of the mobility observed at present is not related to such flows; usually only patients with higher education/higher income ask for these treatments because they require a higher level of health literacy. Our model, where interpersonal income redistribution cannot be modelled, is not suitable for this analysis, but some considerations may be inferred using the results presented in Brekke et al. (2011). When only a fraction of patients are open to travel abroad because of language and other barriers, the Region form which the mobility flows originates tends to decrease less the quality of their care. In this case distributional effects are more likely to exists (individuals with comparatively low education and income are more likely to fall into the "immobile" part of the population), but on the other hand the level of quality of health care is going to be comparatively higher.

4 Long-Run Considerations

In the analysis proposed in the previous section we have assumed that c_j, the number of patients that in each Region are eligible to receive health care, is set outside the model and depends on cost effectiveness measures that are based on the cost to provide care at local level. In this section we discuss the possible effects that mobility may have on this choice. Opening to cross-border patients implies that in each Region there might be two prices for health care: the production cost and the transfer price. In this context the choice of the internal cost to set cost effectiveness threshold is not so straightforward. In fact in Levaggi and Menoncin (2011), in a context where there are no mobility costs, it is shown that the cost effectiveness threshold depends on the transfer price rather than the internal cost. This is because the latter represents in both Regions the true marginal cost: for the Region that receives patients it is an opportunity cost since it represents the money foregone to treat a resident; for the sending Region it represents the cost to treat an extra patient. Furthermore, given the separation of purchasing and providing functions in most countries, the (possibly) higher price paid to treat patients may become extra profit for the hospital rather than increased consumer surplus.

If this is the case we may expect a trade-off between short and long run effects. For example, the most efficient Region may have an interest in asking a price higher than its marginal cost v_0 in order to increase it welfare. However, this may lead in the long run to a change in the cost/effectiveness ratio and fewer local patients will be treated. On the other hand, the less efficient local authority may be better off in the short-run if a low price is charged, but this may imply an increase in the demand for health care in the future that will be met only partially abroad at a price v_0. These considerations may lead both Regions to set the price for mobility to a level that maximizes their long run utility.

5 Conclusions

During the last decades cross-border health care has gained increasing attention in industrialized countries. This reflects structural, cultural, economic and political changes. An increasing emphasis on patient rights and patient choice in many European countries has meant a surge in the number of people that seek treatment abroad. These trends have contributed to an increased interest on the part of policy-makers in cross-border patient mobility. In general, patient choice is sought as a means to improve efficiency and to enhance quality. Brekke et al. (2014) show that in general this is the case both from a theoretical and an empirical point of view. However, more competition may not necessarily imply an improvement in welfare and when such competition involves contracts among different States the gain from such competition may not be equally shared.[5]

The first problem that the EU had to settle was related to the legal framework for cross border shopping. In this respect several sentences by the European Court of Justice and the Directive on cross border mobility have solved a number of

[5] For a discussion on the implications of integration on the health of Europeans see Mackenbach et al. (2013) and references therein.

problems, but on the other hand they create new litigations since they usually need to be interpreted and they are designed to solve a specific controversy (Greer et al. 2013). In spite of its importance the theme has received little attention from an economic point of view. In this article we have chosen to examine from a welfare point of view two types of patient mobility that are especially relevant for Europe, namely the decision of a member state to send its patients abroad (Regulated Mobility) and the autonomous decision of the patients to seek health care in other member states (Patients Mobility). We show that one of the variables that will determine the welfare implications of cross border patient mobility is the transfer price, both in a "static" framework where the number of patients that have access to health care in each Region depends on the cost to produce health care locally and a "long-run" framework where the transfer price may be used to determine the number of people that receive care. We show that although patient choice improves welfare if we compare it with the no mobility case, the differential effect with respect to patient mobility is quite different. In the short run, if the price for mobility is approaching its maximum level (v_1) the less efficient local authority is better off by allowing patients' choice. In this case in fact the extra payment made to local authority zero will be partially used to compete with one in terms of quality. In our model quality is a sort of public good because all the patients receiving care in the same Region receive the same level of care, independently of the number of patients treated. This means that the increased quality will cause a positive externality also on the less efficient local authority. Finally it is interesting to note that welfare improves also for the more efficient Region that is probably using the extra resources in a more productive way. However, in the long run the use of a high reimbursement for mobility may lead to a decrease in total welfare.

A.1 Appendix

A.1.1 Derivation of the Optimal Quantities

1. *No Mobility*
   The optimal quality $q_{j,/;NM}$ is derived from the maximization of (5) over q_j. The F.O.C. gives the condition: $-\theta_j q_j + \frac{1}{2}(1 - c_j) = 0$, which gives the result in (6).
2. *Regulated mobility*
   In this case for Region 0 the welfare in (10) has to be maximized for r and q_0. The function is strictly increasing in the first variable, while the derivative with respect to the second one is the same as for the NM case. Then Region 0 will accept any number of patients as long as $p \geq v_0$ without changing the offered quality.

For Region 1 the welfare function changes and can be written as in (12) and the maximization is done on r and q_1. The F.O.C. gives the following conditions:

$$\begin{cases} (v_1 - p + q_0 - q_1 + m(1-2r))(1-c_1) = 0, \\ -\theta_1 q_1 + (1-c_1)(1-r) = 0. \end{cases}$$

The solutions of the above linear system are reported in (14) and (15). In order for them to be feasible the condition $\frac{1}{2} < r_{RM} < 1$ has to be satisfied. This implies that the following condition has to be verified

$$0 < \frac{q_{0,NM} - m + v_1 - p}{1 - c_1 - 2m\theta_1} < \frac{1}{2\theta_1}.$$

From the second inequality we get

$$\frac{q_{0,NM} + v_1 - p}{1 - c_1 - 2m\theta_1} < q_{1,NM} \frac{1}{1 - c_1 - 2m\theta_1}$$

and since $q_{0,NM} + v_1 - p > q_{1,NM}$ this is possible only if $1 - c_1 - 2m\theta_1 < 0$, which further entails the condition $q_{0,NM} - m + v_1 - p < 0$. The given optimal value for r is then feasible only if

$$m > \max\left(\frac{1-c_1}{2\theta_1}, \frac{1-c_0}{2\theta_0} + v_1 - p\right).$$

3. *Patient choice*

Region 0 maximizes (19) over q_0; the F.O.C. in this case is

$$-\frac{1}{2m}(v_0 - p)(1 - c_1) - \theta_0 q_0 + \frac{1}{2}(1 - c_0) = 0,$$

which does not depend on the quality chosen in Region 1 and gives the optimal quality $q_{0,PC}$ given in (20).

Region 1 maximizes (21) over q_1; the F.O.C. is

$$-\theta_1 q_1 + \frac{1-c_1}{2m}(q_1 - q_0 - v_1 + p + m) = 0$$

and the optimal quality is thus given by (22).

The number of moving patients is given by

$$\begin{aligned} z &= \frac{1}{2} + \frac{1}{2m}\left(q_{0,PC} - q_{1,PC}\right) \\ &= \frac{1}{2} + q_{0,NM}\frac{\theta_1}{-1 + c_1 + 2m\theta_1} + \frac{1}{2}\frac{(1-c_1)(\theta_1(p - v_0) + \theta_0(v_1 - p - m))}{(-1 + c_1 + 2m\theta_1)\theta_0 m} \end{aligned}$$

which is greater than $\frac{1}{2}$ because $q_{0,PC} \geq q_{0,NM} > q_{1,NM} > q_{1,PC}$ and under (13) is also less than 1. In fact $z<1$ iff $q_{0,PC} - q_{1,PC} < m$ and substituting in the inequalities the values in (20) and (22) the condition is equivalent to require that

$$\frac{2m^2\theta_1\theta_0 - (1-c_0)\theta_1 m - (\theta_0(v_1-p) - \theta_1(v_0-p))(1-c_1)}{2m\theta_1\theta_0} > 0.$$

The ensuing condition

$$m \geq \frac{1-c_0}{4\theta_0} + \frac{1}{4}\sqrt{\frac{(1-c_0)^2}{\theta_0^2} + 8\left(\frac{v_1-p}{\theta_1} + \frac{p-v_0}{\theta_0}\right)(1-c_1)} \tag{27}$$

is always verified under hypothesis (13) because

$$\frac{1-c_0}{2\theta_0} + v_1 - v_0 > \frac{1-c_0}{4\theta_0} + \frac{1}{4}\sqrt{\frac{(1-c_0)^2}{\theta_0^2} + 8\frac{v_1-v_0}{\theta_0}(1-c_1)}$$

and since $\theta_0 < \theta_1$ the right-hand side is greater than the lower bound in (27).

A.1.2 Welfare Analysis for the Patient Choice Case

The welfare difference between the PC and the RM case is given in Eq. (25) and by substitution of the optimal values $q_{0,PC}$, $q_{1,PC}$ and r_{RM} it can be written as

$$\begin{aligned}\frac{1}{8}\frac{(1-c_1)(p-v_0)}{m^2\theta_0(-1+c_1+2m\theta_1)}\Bigg[&\left(8\theta_0 m^2\theta_1 - 4m(1-c_1)\left(\theta_0 - \frac{1}{2}\theta_1\right) + (1-c_1)^2\right)p\\ &- 8\theta_0 m^2\theta_1 v_1 + (4(1-c_1))\left(\theta_0 - \frac{1}{2}\theta_1\right)v_1 m\\ &- (1-c_1)^2 v_0 + 2(1-c_1))\theta_1(v_1-v_0)m\Bigg].\end{aligned}$$

The term outside the square bracket is positive and by algebraic tools it is easy to show that the term multiplying p inside the square bracket is positive for $\theta_0 < \theta_1$. Thus the welfare difference is positive whenever p is greater than the bound in (26).

For Region 1 the best way to proceed is to analyze the behavior wrt p of the two optimal welfare levels $W_{1,PC}$ and $W_{1,RM}$ obtained by substitution of the optimal quantities in (21) and (12) resp. Since

$$\frac{\partial^2 W_{1,RM}}{\partial p^2} = \frac{(1-c_1)\theta_1}{-1+c_1+2m\theta_1}$$

$W_{1,RM}$ is convex in p. As for $W_{1,PC}$ we have:

$$\frac{\partial^2 W_{1,PC}}{\partial p^2} = \frac{(1-c_1)^2}{(-1+c_1+2m\theta_1)^2}\left(2\theta_1 AB - \theta_1 B^2 - \frac{1-c_1}{2m}A^2\right)$$

with

$$A = \frac{\theta_1 - \theta_0}{\theta_0}, \qquad B = -1 + \frac{1-c_1}{2m\theta_0}.$$

The term A is always positive, while $B > 0$ iff $m < \frac{1-c_1}{2\theta_0}$. Since $1 - c_0 > 1 - c_1$, by (13) $m > \frac{1-c_0}{2\theta_0} + v_1 - v_0 > \frac{1-c_1}{2\theta_0}$, thus B is always negative and $W_{1,PC}$ is concave. Thus the difference $W_{1,PC} - W_{1,RM}$ is concave. Since for $p = v_0$ we have $W_{1,PC} < W_{1,RM}$, while for $p = v_1$ it is $W_{1,PC} > W_{1,RM}$, it can be concluded that for $p \in (v_0, v_1)$ a unique value p^* exists and such that for $p < p^*$ the welfare level in RM is greater than in PC, while the contrary is true for $p > p^*$.

The welfare difference for Region 1 between PC and NM is given by

$$\begin{aligned} W_{1,PC} - W_{1,NM} = \; & (v_1 - p)(1-c_1)\left(z - \frac{1}{2}\right) - \frac{1}{2}\theta_1\left(q_{1,PC}^2 - q_{1,RM}^2\right) \\ & + (1-c_1)\left(q_{0,PC}\left(z - \frac{1}{2}\right) - \frac{1}{2}m\left(z^2 - \frac{1}{4}\right) + q_{1,PC}(1-z)\,.\right. \\ & \left. -\frac{1}{2}m(1-z)^2 - \frac{1}{2}q_{1,RM} + \frac{1}{8}m\right). \end{aligned}$$

As shown above, $W_{1,PC}$ is concave. Also we have

$$\begin{aligned} \left(\frac{\partial W_{1,PC}}{\partial p}\right)_{p=v_0} = \; & \frac{1-c_1}{4\theta_0^2 m(2\theta_1 m - 1 + c_1)}\left(2\theta_0^2\theta_1\left(-\frac{1-c_0}{\theta_0} + \frac{1-c_1}{\theta_1}\right)m\right. \\ & \left. - (1-c_1)(-\theta_0 c_1 - 2\theta_0^2 v_0 + \theta_1 c_0 + 2v_1\theta_0^2\theta_1 + \theta_0 - 2v_1\theta_0\theta_1 + 2\theta_0 v_0\theta_1)\right) \end{aligned}$$

which, since $-\frac{1-c_0}{\theta_0} + \frac{1-c_1}{\theta_1} < 0$, is negative if

$$m > \frac{1-c_1}{2\theta_0}\left(\frac{2(v_1 - v_0)\,(\theta_0 - \theta_1)}{\theta_1\left(-\frac{1-c_0}{\theta_0} + \frac{1-c_1}{\theta_1}\right)} + 1\right)$$

and by (13) this condition is always verified. Thus we can conclude that $W_{1,PC}$ is strictly decreasing. Since

$$\begin{aligned} (W_{1,PC} - W_{1,NM})|_{p=v_1} = \; & \frac{1-c_1}{8\theta_0^2\theta_1 m^2(2m\theta_1 - 1 + c_1)} \\ & \cdot (2m\theta_1 q_{0,NM}\theta_0 + \theta_1 v_1 - v_0\theta_1 - \theta_0 m - \theta_1 c_1 v_1 + c_1 v_0\theta_1 + \theta_0 m c_1)^2 \end{aligned}$$

is positive, it holds that $W_{1,PC} > W_{1,NM}$ for all $p \in [v_0, v_1]$.

References

Alleman, B. W., Luger, T., Schacht Reisinger, H., Martin, R., Horowitz, M. D., & Cram, P. (2011). Medical tourism services available to residents of the United States. *Journal of General Internal Medicine*, *26*(5), 492–497.

Brekke, K. R., Cellini, R., Siciliani, L., & Straume, O. R. (2012). Competition in regulated markets with sluggish beliefs about quality. *Journal of Economics and Management Strategy*, *21*(1), 131–178.

Brekke, K. R., Gravelle, H., Siciliani, L., & Straume, O. R. (2014). Patient choice, mobility and competition among health care providers. In R. Levaggi & M. Montefiori (Ed.), *Health care provision and patient mobility: Health integration in the European Union*. Italia: Springer-Verlag.

Brekke, K. R., Levaggi, R., Siciliani, L., & Straume, O. R. (2011). Patient mobility, health care quality and welfare. CEPR Discussion Papers 8559, C.E.P.R. Discussion Papers.

Burge, P., Devlin, N., Appleby, J., Rohr, C., & Grant, J. (2004). Do patients always prefer quicker treatment? A discrete choice analysis of patients' stated preferences in the London patient choice project. *Applied Health Economics and Health Policy*, *3*(4), 183–194.

Glinos, I., Baeten, R., & Boffin, N. (2006). Cross-border contracted care in Belgian hospitals, in patient mobility in the European Union: Learning from experience. *European Observatory on Health Systems and Policies*, Copenhagen, pp. 97–118.

Glinos, I. A., Baeten, R., & Maarse, H. (2010). Purchasing health services abroad: Practices of cross-border contracting and patient mobility in six European countries. *Health Policy*, *95*(2–3), 103–112.

Greer, S. L., Hervey, T. K., Mackenbach, J. P., & McKee, M. (2013). Health law and policy in the European Union. *The Lancet*, *381*(9872), 1135–1144.

Legido-Quigley, H., Glinos, I., Baeten, R., & McKee, M. (2007). Patient mobility in the European Union. *British Medical Journal* *334*(7586), 188–190.

Levaggi, R., & Menoncin, F. (2011). Decentralisation in the provision of merit and impure public goods: Who gains, who loses. *SSRN eLibrary*.

Mackenbach, J. P., Karanikolos, M., & McKee, M. (2013). The unequal health of Europeans: Successes and failures of policies. *Lancet*, *381*(9872), 1125–1134.

Mainil, T., Van Loon, F., Dinnie, K., Botterill, D., Platenkamp, V., & Meulemans, H. (2012). Transnational health care: From a global terminology towards transnational health region development. *Health Policy*, *108*(1), 37–44.

Santuari, A. (2013). The European Union Directive on the application of patients' right in cross-border health care. *Journal of Public Health Research*, *2*(1).

Wagner, C., & Linder, R. (2010). The demand for EU cross-border care: An empirical analysis. *Journal of Management and Marketing in Healthcare*, *3*(2), 176–187.

Wildasin, D. E. (2001). Externalities and bailouts: hard and soft budget constraints in intergovernmental fiscal relations. Public Economics 0112002, EconWPA.

Wildasin, D. E. (2004). The institutions of federalism: Towards an analytical framework. *National Tax Journal*, *LVII*, 247–272.

Quality Competition and Uncertainty in a Horizontally Differentiated Hospital Market

Marcello Montefiori

Abstract The chapter studies hospital competition in a spatially differentiated market in which patient demand reflects the quality/distance mix that maximizes their utility. Treatment is free at the point of use and patients freely choose the provider which best fits their expectations. Hospitals might have asymmetric objectives and costs, however they are reimbursed using a uniform prospective payment. The chapter provides different equilibrium outcomes, under perfect and asymmetric information. The results show that asymmetric costs, in the case where hospitals are profit maximizers, allow for a social welfare and quality improvement. On the other hand, the presence of a publicly managed hospital which pursues the objective of quality maximization is able to ensure a higher level of quality, patient surplus and welfare. However, the extent of this outcome might be considerably reduced when high levels of public hospital inefficiency are detectable. Finally, the negative consequences caused by the presence of asymmetric information are highlighted in the different scenarios of ownership/objectives and costs. The setting adopted in the model aims at describing the up-coming European market for secondary health care, focusing on hospital behavior and it is intended to help the policy-maker in understanding real world dynamics.

Keywords Competition · Hospitals · Information · Quality

JEL Classification I11 · I18 · L13

M. Montefiori (✉)
Department of Economics, University of Genoa, Via Vivaldi 5 16126 Genoa, Italy
e-mail: montefiori@unige.it

R. Levaggi and M. Montefiori (eds.), *Health Care Provision and Patient Mobility*, Developments in Health Economics and Public Policy 12,
DOI: 10.1007/978-88-470-5480-6_9,

1 Introduction

Patient mobility for medical care is a relatively new phenomenon that continues to grow in importance at national and international level. Patient mobility within national borders is already well established in a number of European countries, even if the organization of the health care market is quite different from country to country. In Italy, the ever greater financial autonomy and devolution enjoyed by the regions (that began in the mid-nineties) has determined the creation of very diverse and independent regional health services. However, it is the patient's right to choose the provider he prefers, both from within and outside the regional borders, from either the public or private[1] providers. Furthermore it has to be noticed that patients do not pay[2] for the health service they receive (even when it is provided by private hospitals) because the national/regional health service is publicly funded by general taxation (Montefiori 2005; Levaggi and Montefiori 2013). This setting, where patients are free to choose the hospital they prefer from public and private ones, is seen in an increasing number of European Countries with public health care systems (Gravelle and Sivey 2010; Montefiori 2005; Montefiori 2008; Appleby and Dixon 2004; Vrangbaek and Ostergren 2006).

There is a similar system in Spain. Hospitals are reimbursed by prospective payment (DRG based) and patients are free to choose the provider they prefer from either the public or accredited private institutions belonging to the "Areas de Salud" ("Health Areas"). In France, patients receive guidance from general practitioners and they choose from public and private health care providers. The hospital's activity is reimbursed by a mixed system which consists of two components: a block grant and a prospective payment.

However, the crucial point on which we want to focus attention is that nowadays, because of the process of globalization, even cross-border health mobility is growing in relevance and importance.

A recent European Directive (Directive 2011/24/EU) of the European Parliament states the right of EU patients to be financially covered in the event that they go to another country to receive healthcare services, under the condition that they were also offered it by their own health system (Brekke et al. 2012b). However there is still a lot of discretion left to member states to establish the rules and terms of reimbursement, and the problem remains indeed unsolved.

"While citizens in the EU, in principle, are free to seek health care wherever they want and from whatever provider available, in practice this freedom is limited by their ability to pay for it or by the conditions set out by public and private funding systems for health care" (Palm and Glinos 2010).

---

[1] To note that a private hospital that wishes to be "accredited" has to meet specific requirements set by the national and regional government. In particular, private hospitals that want to work for the public sector are required to meet the quality standard and accept the same prospective reimbursement (DRG based) provided for public hospitals (Levaggi and Montefiori 2013).

[2] But a possible co-payment might be required from the patient in some circumstances.

The goal of patient mobility is subject to the conditions of international agreements between countries. If patients were free to choose the provider they prefer among those within the EU and the service was still free at the point of use (for example, by defining a sort of "mobility DRG tariff" to compensate for the inwards/outwards movement of patients) then a quality-competitive market would take place at EU level, with positive outcome both in terms of efficiency and quality (Chalkley and Malcomson 1998a, b; Gravelle 1999; Gravelle and Masiero 2000, 2002; Ma 1994; Montefiori 2008).

Hence we are going in the direction of an integrated international health care market[3] in which hospitals that differ from each other in terms of costs, characteristics and objectives compete for patients. However, at the same time, in this broadening market also the demand (i.e., the current patient characteristics) may differ noticeably depending on the country of origin.

In particular, the perception and expectation of the quality level provided by the hospital might be differently biased: because of the asymmetry of information that characterizes the health care market, the actual quality provided by the hospital might be incorrectly observed in a stochastic framework (Gravelle and Masiero 2000, 2002; Montefiori 2005, 2008). It is evident that uncertainty plays a crucial role in conditioning hospital behavior.

This chapter aims to investigate different scenarios in which hospitals compete for patients in a spatially differentiated market. In particular, the issues of uncertainty and asymmetric objectives will be jointly considered. The model that will be used assumes that hospitals are paid by prospective payments that consist of a fixed price per treated patient (Levaggi 2005, 2007; Montefiori 2008). The price is set as the average cost incurred by hospitals when treating a patient with a specific diagnosis. In addition, it is assumed that patients do not pay for the health services they receive since they are obtained free of charge at the point of use.

The afore-mentioned setting should create a quality-competitive hospital market, i.e., a market in which providers compete on quality in order to attract patients. In fact, since the health service is free at the point of use and the price is fixed, the only means at a hospital's disposal to increase the number of patients (and the revenue) is to invest on quality. Note that the same result is expected to be obtained in the case where all patients are assumed to be insured or when a copayment is required (both in the case of a tax financed health care system and of a private insurance) under the condition that copayments do not vary among hospitals for the same treatment/diagnosis.

In order to analyze the quality-competitive, horizontally differentiated market, the basic model presented in Montefiori (2005) is used and extended.

The model in Montefiori (2005) focuses on the effects of competition between two hospitals that are symmetric in terms of costs and objectives. Hospitals are profit maximizers and compete for patients on quality in a Hotelling type spatially

[3] In general, patient mobility within the EU is negligible, however there are exceptions of countries and regions that cope with high level of mobility flows (Palm and Glinos 2010).

differentiated market (Hotelling 1929). That paper introduced the new approach of mean–variance in order to take into account the problem of uncertainty. This biases the patient's perception of the quality level provided by hospitals for medical services.

However the study presented in this paper differs considerably from Montefiori (2005) because of a very different setting in terms of hospital (asymmetric) objectives and differentiation in terms of costs. Furthermore, a new type of uncertainty is proposed and its consequences for the market equilibrium outcome are analyzed.

The present study aims to analyze the effects of effective competition on quality among providers from the theoretical point of view and by numerical simulations. The opinion of the author is that the model used in this paper fits well the current/forthcoming scenario where hospitals compete for patients within a macro-area such as the EU. However it also applies to a "within-the-home state" competition scenario.

The chapter will be organised as follows: in Sect. 2 the model is presented. The subsequent Sect. 3 provides the analysis of the equilibrium when a simultaneous Nash–Cournot competition takes place. Section 4 presents the sequential Stackelberg quality choice equilibrium in the case when hospitals have symmetric objectives but asymmetric costs. The implications of uncertainty and the definition of "perceived quality", which might be biased with respect to the actual quality, are introduced in Sect. 5. Numerical simulations referring to all the scenarios previously studied from the theoretical point of view are reported in Sect. 6. Finally Sect. 7 summarizes the main conclusions.

2 The Model

The purpose of this paper is the study of hospital health care using a Hotelling-spatial-competition model. A simple linear Hotelling-type market is considered. A mass of patients (for simplicity normalized to 1) is uniformly distributed on a unit-length line market. Each patient demands only one medical treatment. It is assumed that the market is served by only two hospitals ($i = A, B$) whose location is exogenously set at the two extremes of the line.

In order to analyze the quality-competitive, horizontally-differentiated market, the basic model presented in Montefiori 2005 is used and extended.

2.1 Hospitals

Two distinct hospitals compete for patients and they might pursue symmetric or asymmetric objectives. When the two hospitals are both private or public, then it is assumed that they have symmetric objectives. However when a private hospital competes with a publicly owned one, then a mixed market with asymmetric objectives takes place. For private hospitals profit maximization is generally the most common assumption. Public hospitals are, in contrast, heterogeneous in their objectives. In this work it is assumed that the public hospitals may pursue alternatively two different objectives: profit maximization and quality maximization. The latter reflects, to some extent, one of the scenarios suggested in Levaggi and Montefiori (2013) under the behavior that the authors name as "excellence" but with the striking difference that in the present paper the budget constraint is hard (both for the public and the private hospital).

The hospital receives a prospective reimbursement M for each patient. The number of patients depends on the demand D which in turn depends on the quality differential between the two hospitals

The hospital cost function depends on: (i) the number of patients (i.e., the demand D); (ii) the hospital specific cost parameter c_i; (iii) the quality level provided. For simplicity and without loss of generality the fixed costs are set to zero.

Therefore the hospital profit function can be written as:

$$\pi_i = [M - c_i q_i] D_i. \quad (1)$$

The hospital participation constraint is met when the purchaser is able to set a contract that grants the hospital (at least) its reservation profit which is set equal to zero. In our model this implies: $M \geq c_i \bar{q}_i$ where $\bar{q}_i > 0$ is a given level of quality exogenously set by the regulator and that might be different from that which the hospital sets.

The only choice variable for the hospital is its own quality level.

2.2 Patients

Patients aim at utility maximization. This is positively affected by quality and negatively affected by distance. As already stated, patients receive the medical care they require free of charge at the point of use since a public health care system funded by general taxation is assumed. Nonetheless, patients face traveling costs related to the distance of the health care provider. The key elements that causes stiffness in healthcare mobility are, in fact, the monetary costs (but also the non-monetary ones) related to mobility that are not covered by the national health system. A high quality level provided by hospitals located far away, might motivate patients to meet monetary and non-monetary costs inherent in traveling for care (Levaggi and Montefiori 2013; Montefiori 2005; Sanjo 2009). The non-monetary costs are generally related to the patients' preference in receiving care close to home (and possibly in one's own country) where relatives are able to come to visit and patients themselves feel comfortable with an environment they know (Montefiori 2005).

Quality and distance affect the utility function by means of the α and the γ parameters respectively. The γ parameter synthesizes rigidity in patient mobility which is mainly due to individual preferences (Brekke et al. 2011b; Montefiori 2005). This, for example, might be determined because of cultural and language differences among countries.[4]

$$U = v + \alpha q_i - \gamma d_i; \quad i = a,\ b \tag{2}$$

where q_i is the quality level provided by hospital i: the higher the quality the higher is the patient's utility; d_i is the patient distance from hospital i: the greater the distance from the hospital the lower is the patient's utility; v is the valuation of the treatment (Brekke et al. 2006, 2011a, b; Herr 2011) and it is assumed to be high enough to meet the patient's participation constraint even when the quality level set by the hospital is set equal to zero. In this paper the minimum quality level that the hospitals are forced to provide in order to avoid malpractice is assumed to be zero (Chalkley and Malcomson 1998a, b; Montefiori 2008). These assumptions ensure full market coverage.

The location of the marginal patient, that is the patient who is indifferent between hospital i and hospital j, is: $d_i = \frac{\alpha}{2}\left(q_i - q_j\right) + \frac{1}{2};\ \ i,\ j = a,\ b;\ i \neq j$.

Because the mass of patients uniformly distributed on the unit-length line market is normalized to 1, then the location of the marginal patient d_i is also the demand D_i for hospital $(i = a,\ b)$.

2.3 Social Welfare

In this chapter we are also interested in evaluating the equilibriums in light of the effects on welfare. Coherently with the existing literature, social welfare is considered as the sum of individuals' payoffs. Henceforth two addendum have to be jointly considered: the hospital and the consumer surpluses.

The consumer surplus, knowing that patients are uniformly distributed on the unit-length line market and that the market is fully covered, is given by the sum of the surplus referred to those receiving the treatment from hospital A and the surplus of those receiving treatment from B:

$$S_i = \int_0^{D_i} U(q_i,\ l_i)\, dl_i \tag{3}$$

where S_i is the overall surplus attained from patients receiving medical care from hospital i $(i = a,\ b)$.

4 In some cases the disutility in distance γ is interpreted in relation to the utility in quality α (Brekke et al. 2012a), i.e., by normalizing α to 1 and interpreting γ as a relative marginal disutility.

Note that, for sake of clarity, the letter *d* in Eq. 2 (referring to patient's distance from hospital) has been substituted with the letter *l* (maintaining exactly the same meaning of *d*) in order to avoid chaos in the formula for integrals.

Using Eq. 1 and summing it up with 3, the social welfare *W* can then be defined as:

$$W = \sum\nolimits_{i=a}^{b} \pi_i + \sum\nolimits_{i=a}^{b} S_i. \tag{4}$$

3 The Simultaneous Nash–Cournot Equilibrium

3.1 Benchmark

In this section the hypothesis of two hospitals with symmetric costs and symmetric objectives, that compete in a context of perfect information, is assumed. This scenario is intended to be the benchmark[5] to which we compare the next ones.

The hospitals aim at profit maximization. This is the case when two private hospitals compete for patients, or, alternatively, when the public hospitals pursue the profit maximization goal.

Hospitals behave strategically in a non-cooperative game by setting the quality level that maximizes the profit. Because hospitals perfectly know other hospitals best reply, they set their quality taking into account the competitor's reaction. The quality setting is provided in Eq. 5

$$q_i = \frac{M}{c} - \frac{\gamma}{\alpha}; \quad i = a,\ b. \tag{5}$$

3.2 Symmetric Objectives but Asymmetric Costs

In this scenario it is assumed that both hospitals maximize their profits but they face different marginal production costs. The context is still that of perfect information. Again, the non-cooperative game is solved to identify the Nash-equilibrium quality level

$$q_i = \frac{M}{3}\left(\frac{2}{c_i} + \frac{1}{c_j}\right) - \frac{\gamma}{\alpha}. \tag{6}$$

Substituting Eq. 6 in 1, the Nash equilibrium hospital's profit can be identified:

[5] The benchmark coincides with the results provided in Montefiori (2005).

$$\pi_i = \frac{\left(M\alpha c_i - M\alpha c_j - 3\gamma c_i c_j\right)^2}{18 c_i c_j^2 \alpha\gamma}; \quad i,j = a,\ b;\ i \neq j. \tag{7}$$

Looking at Eq. 7 it clearly emerges that when hospitals face the same cost c (i.e., $c_a = c_b = c$) then the hospitals provide the same quality level (Eq. 5) and get the same level of profit $\frac{c\gamma}{2\alpha}$.[6]

Assuming that the hospital's participation constraint is met and that $c_i < c_j \leq M$, then the equilibrium level for quality will be $q_i > q_j$. The market share is then given by:

$$D_i = \frac{1}{2} - \frac{\Omega}{6}; \quad \mathrm{i = a,\ b}$$

where:

$$\Omega = \frac{M\alpha\left(c_i - c_j\right)}{\gamma c_i c_j}; \text{ and } -3 \leq \Omega \leq 3; \ \mathrm{i,j = a,\ b;\ i \neq j.} \tag{8}$$

Ω assumes value zero when $c_i = c_j$; the two hospitals will equally share the market.

Ω will be greater than zero when $c_i > c_j$ and lower than zero when $c_j > c_i$.

The more efficient hospital will be able to get a larger share of the market by the higher quality level it will be able to provide with respect to the competitor. As a consequence, higher profits are expected.

3.3 Asymmetric Objectives and Asymmetric Costs

It is possible that the two hospitals have asymmetric objectives. In fact, if, on the one hand, profit maximization is a reasonable assumption for private hospitals, on the other, public hospitals may pursue objectives different from profit maximization, such as social welfare or reputation (see Levaggi and Montefiori (2013) for details), In this section it is assumed that the public hospital A aims at quality maximization (under the assumption of hard budget constraint) while the private hospital B still aims at profit maximization. This setting reflects to some extent the "excellence" case provided in Levaggi and Montefiori (2013) but with the striking difference that, in their model, the public hospital interprets its budget as soft and that the regulator systemically bails out the public hospital deficit.

Henceforth in this section it is assumed that the public hospital aims at quality maximization under the assumption of hard budget constraint. In order to avoid negative profit, hospital A takes into account hospital B's best reply, and sets the

---

[6] See Montefiori (2005) for details.

maximum quality that can be provided, avoiding the risk of incurring negative profits. In doing so the public hospital's quality setting is:

$$q_a = \frac{M}{c_a}. \tag{9}$$

The hospital B quality level can then be written as:

$$q_b = \frac{1}{2}\left(\frac{M}{c_a} + \frac{M}{c_b} - \frac{\gamma}{\alpha}\right). \tag{10}$$

From Eq. 10 it is possible to derive the market share served by each hospital:

$$D_a = \frac{3}{4} - \frac{\Omega}{4}; \; D_b = \frac{13}{4} + \frac{\Omega}{4}. \tag{11}$$

Equation 10 suggests that in most cases hospital A will provide higher quality with respect to the private hospital B. This implies that when the two hospitals have the asymmetric objectives presented in this scenario but the same marginal cost c then the public hospital market share will be 75 % of the entire market. However, this share might be reduced when the public hospital is inefficient, i.e., when $c_a < c_b$. In fact, because of public hospital inefficiencies, it could be the case that q_b was not lower than q_a. In particular this applies when the following condition is met:

$$c_a \geq \frac{Mc_b\alpha}{M\alpha - \gamma c_b}. \tag{12}$$

It is straightforward to verify that, when condition 12 holds with equality, the two hospitals provide the same quality level[7] and, as a consequence, they equally share the market.

4 The Sequential Stackelberg Equilibrium: Symmetric Objectives but Asymmetric Costs

When hospitals differ in management or size, a sequential game might take place. Assuming hospital A to be public (or a large hospital) and B private (or a small hospital), then one might expect that the former was the leader while the latter the follower, in a sort of Stackelberg sequential game.

In this section it is assumed that both hospitals, regardless of their ownership, maximize profits and it is assumed that they face different marginal costs (due to their dimensional structure, for instance).

Hospitals share the market with reference to a single DRG.

---

[7] $q = \frac{M}{c} - \frac{\gamma}{\alpha}$.

The leader tries to take advantage of the fact that it is the first mover. However this advantage does not result in a better outcome with respect to the competitor's. Let's see the timing of the game.

The leader moves first taking the follower's best reply into account. The follower observes the quality level set by the leader and, in turn, sets its profit maximizing quality level.

The equilibrium quality settings for hospital A and hospital B will turn out to be:

$$q_a = \frac{M}{2}\left(\frac{c_a + c_b}{c_a c_b}\right) - \frac{3\gamma}{2\alpha} \tag{13}$$

$$q_b = \frac{M}{4}\left(\frac{c_a + c_b}{c_a c_b}\right) + \frac{M}{2c_b} - \frac{5\gamma}{4\alpha} \tag{14}$$

while the market share is, respectively:

$$D_a = \frac{3}{8} - \frac{\Omega}{8}; \; D_b = \frac{5}{8} + \frac{\Omega}{8}. \tag{15}$$

Equations (13) and (14) suggest that for any $c_a > \frac{M\alpha c_b}{M\alpha + \gamma c_b}$ the quality provided by hospital B will be greater than the quality provided by A ($q_a < q_b$).

The demand function of each hospital reflects the quality level provided.

It is straightforward to grasp from the conditions above that, under the assumption that the two hospitals face the same marginal cost ($c_a = c_b$), the leader turns out to be the loser. The counterintuitive outcome obtained, comes from the two-stage game we have defined. The hospitals compete on quality, given the fixed price M per treated patient. The hospital which sets its quality level first loses, even when it takes the competitor's reaction function into account. In fact, the follower observes the leader setting its quality and, only at a second stage, sets its quality level in order to maximize its profit. The hospital will find it profitable to set a slightly higher quality with respect to the rival. By this behavior the follower gets a larger share of the market ($D_b > D_a$) and in so doing it increases its profit. Summing up, in the case that both hospitals face the same costs and have the same revenue function, the second mover wins. Few authors, with reference to industrial economics, have found something similar. For instance Beato and Mas-Colell (1984) find, under specific assumptions, the same odd result. However, in the long-run, the interesting result is that the sequential equilibrium converges towards the simultaneous equilibrium outcome that we have identified in the previous sections.[8]

---

[8] The two-stage sequential competition on quality can be extended dynamically. The equilibrium outcome found is not actually dynamically stable. The hospitals will move from it in subsequent stages of the game. If we preserve the game structure of the equilibrium, we can assume the leader will be able to react to the follower's behavior in the next stage of the game. The leader will respond to the follower's quality, maximizing a new objective function where the

5 Uncertainty

Up to this point, perfect observation of quality level, by patients, has been assumed. However this assumption is not respondent to the health care market in which it is unrealistic to consider perfect information, at least with reference to patients' ability to observe the actual quality level provided by hospitals. For this reason, in this section the perfect information assumption is relaxed and patients' biased observation of quality is introduced. To cope with this issue the Montefiori (2005) model is extended (refer to it for details) in this section. Patient behavior is affected by the "perceived quality" $\tilde{q}$ which depends on the actual quality and on an error term:

$$\tilde{q}_i(q_i,\ \varepsilon_i) \text{ where } \varepsilon_i \sim N(0,\ \sigma_i^2).$$

Basically, the model suggested in Montefiori (2005) is adopted here but with the striking difference that the error term is characterized by heteroskedasticity; its variance σ^2 varies according to a patient's distance from the hospital.

Patient utility can then be written as[9]:

$$U^i_{pz} = \alpha q_i - \gamma d_i - \beta\sigma_i^2 d_i \tag{16}$$

where $\tilde{q}$ is the perceived quality and σ_i^2 is the variance of the error term weighted by the distance parameter d. When patients are close to the hospital $(d \to 0)$ then the model assumes an error variance which tends to zero. On the other hand, a larger value for d allows for larger uncertainty, represented by an error term variance that tends to σ_i^2 $(d \to 1)$.

Solving the simultaneous equation systems represented by the hospital's stochastic reaction functions, the equilibrium levels for quality are derived (Eq. 17).

$$q_i = \frac{1}{3}\left[M\left(\frac{2}{c_i} + \frac{1}{c_j}\right) - \frac{\beta}{\alpha}\left(\sigma_i^2 + 2\sigma_j^2\right)\right] - \frac{\gamma}{\alpha}. \tag{17}$$

The undesirable consequence in terms of quality deriving from the observation bias is easily detectable from Eq. 17: because of the term $-\frac{\beta}{3\alpha}\left(\sigma_i^2 + 2\sigma_j^2\right)$ that enters the utility function and represents patient uncertainty on quality, the quality equilibrium level provided by hospitals is noticeably reduced. Obviously this turns out to be a gain for the hospitals, in terms of larger profits, and a loss for patients,

(Footnote 8 continued)
competitor's quality is given. In this way we define a third stage new equilibrium. This equilibrium will change again in the fourth stage when the follower will move after the observation of the leader's quality. Unavoidably the long-run equilibrium will converge to that already found in the simultaneous quality choice. Hospitals' long-run dynamic competition converges towards the simultaneous Nash equilibrium. In the long-run hospitals competing in a multi stage game produce the same result as in the simultaneous equilibrium.

[9] See Montefiori (2005) for details and required mathematical steps.

in terms of a reduction in their surplus (this statement will be extended in the simulation section).

6 Simulations

In this section numerical simulations for the scenarios previously investigated are provided. This part aims at providing numerical values to help in grasping information about the equilibrium outcome obtained in the different scenarios.

However, before starting with the numerical analysis, there are general assumptions that have to be set to implement the analysis. All the assumptions declared herewith will remain unchanged for all the simulations presented hereafter, in order to render the outcomes comparable to each other.

Settings are defined in order to render as simple as possible the simulation, but trying to avoid any loss in generality. Assumptions concern the parameters α, γ and *M*. In particular, the parameter α that refers to the marginal utility of quality is set equal to 1; the parameter γ that refers to the marginal disutility of distance is set equal to 1 and the reimbursement *M*, received by hospitals for each patient treated, is set equal to c_a times c_b. In order to maintain the reimbursement *M* constant and equal to μ, the cost c_i is set equal to the ratio $\frac{\mu}{c_j}$, where $i, j = a, b$.

μ is a value exogenously determined and constant.

By these settings we are able to focus on the marginal costs relationship/ratio rather than on the their absolute value (which is not relevant for our purposes). In particular, given an equal value for the reimbursement *M*, the product of hospitals' marginal costs ($c_a \cdot c_b$) stays unchanged and equal to the constant value μ.

In fact, if we allow one hospital to vary its cost, then, from the general equilibrium point of view, it would mean a "general system" resource injection if costs decrease, and in a resource decrease if the opposite occurs.

An alternative hypothesis that has been considered was to set the reimbursement *M* equal to the sum of the marginal costs ($c_i + c_j$). By doing this we were able to isolate the relative cost effect and "compensate" for the "endowment" effect. However, it has to be noticed that even with this different setting, the results found in the simulations wouldn't change in relative terms, while they would necessarily change in absolute value.

6.1 Simulation 1: Symmetric Objectives but Asymmetric Costs

The simulation reported in Table 1 provides the results when the two hospitals aim at profit maximization but they face asymmetric costs. The first column refers to the ratio (c_a/c_b) between the two hospitals' marginal costs.

Table 1 Simulation 1

$\frac{c_a}{c_b}$	q_a	q_b	$\sum_{i=a}^{b} q_i$	π_a	π_b	$\sum_{i=a}^{b} \pi_i$	D_a	D_b	Ω	S_a	S_b	$\sum_{i=a}^{b} S_i$	W
0.7	2.40	2.02	4.43	2.51	0.73	3.24	0.69	0.31	−0.19	1.42	0.58	2.00	5.24
0.8	2.30	2.06	4.36	2.16	1.03	3.19	0.62	0.38	−0.12	1.23	0.72	1.95	5.14
0.9	2.22	2.11	4.33	1.85	1.32	3.17	0.56	0.44	−0.06	1.08	0.84	1.92	5.09
1	2.16	2.16	4.32	1.58	1.58	3.16	0.50	0.50	0.00	0.96	0.96	1.91	5.07
1.1	2.12	2.22	4.33	1.34	1.83	3.17	0.45	0.55	0.05	0.85	1.07	1.92	5.09
1.2	2.08	2.27	4.35	1.13	2.05	3.18	0.40	0.60	0.10	0.76	1.18	1.93	5.12
1.3	2.05	2.33	4.38	0.94	2.26	3.20	0.36	0.64	0.14	0.68	1.28	1.96	5.16

The following factors are reported in the subsequent columns: the quality levels provided by hospital A and hospital B respectively; the overall quality level provided; the individual and overall profits; the demand respectively for hospital A and B expressed in terms of market share (D_a and D_b); the surplus of those consumers who demand medical care from hospital A (S_a); the surplus of those consumers who demand medical care from hospital B (S_b); the overall consumer surplus ($S_a + S_b$); and, in the last column, the social welfare *W*, which is given by the sum of hospital profits and consumer utility.

Looking at Table 1 we can consider the row where the ratio between marginal cost (c_a/c_b) is equal to 1 as the benchmark. This row presents the equilibrium outcome for the different variables of interest when the two hospitals face identical marginal costs. As a consequence, their behavior is symmetric in the sense that they provide the same quality level, serve the same market share and get the same profit.

Moving from the cost equivalence condition (rows above or below the benchmark) we investigate the effects of cost differences between the hospitals. The condition of $c_a/c_b \neq 1$ may also be regarded as a difference in a hospital's efficiency in providing a given level of quality for treating a given number of patients.

Table 1 shows that when the hospital faces lower marginal costs with respect to the competitor, its marginal relative cost for quality decreases and, as a consequence, it has the opportunity to increase its market share by a quality improvement. Remember that it is assumed that one hospital's cost reduction is "compensated" by the other's cost increase in such a way that $c_a\ c_b = \gamma$ and assuming no changes in the reimbursement *M*. To this extent, differences in marginal costs cause, because of the increase in the overall level of quality, the increase in the patients' surplus but also an increase in the hospitals' overall profits (the loss for the less efficient hospital is more than compensated by the gain for the other).

From the social planner perspective the asymmetry in costs between the two hospitals seems to be a desirable goal. In fact the interesting result that emerges from this scenario is that differences in hospitals' costs increase the patient surplus. This comes from the fact that the overall amount of quality tends to increase as the difference in costs increases, assuming the reimbursement *M* stays constant. The competition between the two hospitals, when they differ in efficiency (i.e., in their cost parameter c), incentivises the hospital with lower costs to increase the quality (which turn out to be cheaper for him with respect to the competitor's) in order to serve a larger share of the market and earn extra profits.

This result would be reduced for higher levels of γ ($\gamma > 1$) (where γ is the patients disutility because of distance) but increased for higher values of α ($\alpha > 1$) (where α is the patient utility because of quality).

The hypothesis of different marginal costs reflects the case when large sized hospitals compete with small sized hospitals, the latter characterized by a reduced ability to attract patients by means of quality (Montefiori and Resta 2009). Small sized hospitals with a limited catchment area are useful, from the social welfare perspective, in order to avoid negative utility for patients located nearby who, otherwise, are forced to require medical care from the large sized (high quality) hospitals located far away. In fact, for furthest away patients, the greater utility because of the quality level provided is not sufficient to compensate for disutility because of the distance.

6.2 Asymmetric Objectives and Asymmetric Costs

A mixed market where the two hospitals have asymmetric objectives is assumed here. The public hospital (hospital A) pursues the "*excellence*" goal (see Levaggi and Montefiori 2013) through quality maximization, while the private hospital (hospital B) is still interested in profit maximization.

When the public hospital provides a high level for the quality variable, the resultant consequences on profit are difficult to foresee.

On the one hand, a higher quality increases revenue via the demand mechanism, on the other it increases the costs that depend on the quality itself and on the number of patients. Because of the afore mentioned effects and the hard budget constraint, the public hospital has to take into account the private hospital "best reply" in order to avoid negative profits: the increase in cost has to be balanced by an equivalent increase in revenue.

Looking at Table 2 we can observe, with reference to the symmetric cost case (i.e., when the c_a/c_b ratio is equal to 1), a neat overall quality increase. This outcome is due to the higher quality provided by the public hospital, but also to the higher quality provided by the private one. In fact the latter is forced to react to the high quality provided by the competitor by increasing in turn its quality in order to reduce the loss in terms of profit and contain the market share loss. Comparing the benchmark provided in Table 1 with Table 2 (we have to look at the row with $c_a/c_b = 1$ with reference to both tables) we note that the quality provided by A moves from 2.16 to 3.16, while the quality of B from 2.16 to 2.66. This overall quality increase is obtained at the expense of the hospitals' profit. The public hospital's profit is equal to zero, by the "*excellence*" setting. However, hospital B's profit is positive and equal to 0.4. Note that the benchmark hospital B profit was equal to

Table 2 Simulation 2

$\frac{c_a}{c_b}$	q_a	q_b	$\sum_{i=a}^{b} q_i$	π_a	π_b	$\sum_{i=a}^{b} \pi_i$	D_a	D_b	Ω	S_a	S_b	$\sum_{i=a}^{b} S_i$	W
0.8	3.54	2.68	6.22	0	0.04	0.04	0.93	0.07	−0.71	2.85	0.19	3.04	6.28
0.9	3.33	2.67	6.00	0	0.19	0.19	0.83	0.17	−0.33	2.43	0.43	2.86	6.05
1	3.16	2.66	5.82	0	0.40	0.40	0.75	0.25	0.00	2.09	0.63	2.72	5.89
1.1	3.02	2.67	5.68	0	0.75	0.75	0.67	0.33	0.30	1.81	0.81	2.62	5.78
1.2	2.89	2.68	5.56	0	0.90	0.90	0.61	0.39	0.58	1.56	0.98	2.54	5.71
1.3	2.77	2.69	5.46	0	1.16	1.16	0.54	0.46	0.83	1.36	1.13	2.48	5.67
1.37*	2.70	2.70	5.40	0	1.35	1.35	0.50	0.50	1.00	1.23	1.23	2.03	5.24

* Limiting case where condition (12) is met with equality, i.e., $c_a = \frac{Mc_b\alpha}{M\alpha-\gamma c_b}$

1.58. As previously mentioned, the behavior of hospital B is imposed to contain the loss in terms of market share. In fact, that high level of quality provided by A drastically reduces its demand. To cope with this collapse in the demand, it has to invest in (costly) quality. Nonetheless, the private hospital market share falls from 50 % of the benchmark to 25 % of the present scenario.

Moving to patient surplus, a forgone but substantial increase is obtained. From the social welfare perspective the increase in patient surplus overcomes the profit decrease, moving upwards from 5.07 of the benchmark to 5.89.

Simulation 2 highlights the important result that a publicly managed hospital pursuing an objective different from profit maximization could noticeably improve social welfare. This result would be even more amplified in the case that the publicly managed hospital was characterized by high levels of efficiency (i.e., when the c_a/c_b is lower than 1). The presence of a publicly managed hospital which pursues objectives different from profit maximization would ensure the market with higher level of: patient surplus (S), general welfare (W) and quality (q) with respect to the case of Table 1 where two profit maximizer hospitals compete for patients. Unfortunately publicly managed hospitals are generally characterized by higher levels of inefficiencies with respect to privately owned ones. If the afore- mentioned inefficiencies are symbolized by high values of the marginal cost c, and these inefficiencies reach the threshold value of $c_a/c_b = 1.37$ (the condition of Eq. 12), then the equilibrium outcome would experience a neat curb in terms both of quality and consumer surplus. In the case of very high values of inefficiency ($c_a/c_b >> 1.37$) for the public hospital, the afore mentioned results may change in sign and a socially undesirable outcome may turn out to be the unavoidable consequence.

6.3 Sequential Stackelberg Equilibrium

Here, the scenario presented in Sect. 4 is analyzed by simulations. The two hospitals have symmetric objectives but they do not move simultaneously. In a sequential-Stackelberg game, hospital A, assumed to be the leader, moves first

Table 3 Simulation 3

$\frac{c_a}{c_b}$	q_a	q_b	$\sum_{i=a}^{b} q_i$	π_a	π_b	$\sum_{i=a}^{b} \pi_i$	D_a	D_b	Ω	S_a	S_b	$\sum_{i=a}^{b} S_i$	W
0.7	1.71	1.68	3.39	2.83	1.77	4.59	0.52	0.48	−1.13	0.75	0.69	1.45	6.04
0.8	1.68	1.76	3.44	2.43	2.04	4.47	0.46	0.54	−0.71	0.67	0.80	1.47	5.94
0.9	1.67	1.83	3.50	2.08	2.27	4.35	0.42	0.58	−0.33	0.61	0.90	1.51	5.86
1	1.66	1.91	3.57	1.78	2.47	4.25	0.38	0.63	0.00	0.55	1.00	1.55	5.80
1.1	1.67	1.99	3.66	1.51	2.65	4.16	0.34	0.66	0.30	0.51	1.10	1.61	5.76
1.2	1.68	2.07	3.75	1.27	2.81	4.08	0.30	0.70	0.58	0.46	1.20	1.66	5.74
1.30	1.69	2.15	3.84	1.06	2.95	4.01	0.27	0.73	0.83	0.42	1.30	1.72	5.73

whereas hospital B, which is the follower, moves after observing the competitor's behavior.

The results of the simulation presented in Table 3 show that the advantage of the first move in this kind of sequential-Stackelberg game is not an real advantage.

Looking at the row where the hospitals face symmetric costs ($c_a/c_b = 1$) of Table 3, it is noticeable that the leader (hospital A) gets a lower profit with respect to the follower (hospital B). In other words, we get the surprising result that the leader is the loser. Because of the Stackelberg competition we observe a general quality curb, by which hospitals are able to increase their profits. This produces a general welfare improvement (5.80 instead of 5.07 provided in the benchmark of Table 1) but the latter is obtained at the expense of patients who noticeably reduce their general welfare (from 1.91 of the benchmark to 1.55 in the present scenario).

Still looking at Table 3 it is possible to note that the condition $c_a < \frac{Mc_b\alpha}{M\alpha-\gamma c_b}$(see Sect. 4 for details) which determines $q_a > q_b$ is verified when the ratio c_a/c_b falls below the value 0.8. The policy implication is that a sequential quality choice is able to get the highest level of social welfare when the leader shows a very low marginal cost with respect to the follower. By this condition it is possible to improve the overall amount of profits but still at the expense of the overall amount of patient utility.

Summing up, it is possible to state that when a Stackelberg competition takes place a lower level of quality (and as a consequence a lower level of consumer surplus) is the expected outcome. On the other hand, hospitals (in particular the followers) gain in terms of profit. The Stackelberg competition allows for a social welfare improvement (W). However, only hospitals benefit from this setting (with respect to the benchmark outcome) whereas the consumers are penalized.

6.4 Uncertainty: Symmetric Objectives but Asymmetric Costs

In this section we are interested in understanding better the role played by uncertainty in affecting the equilibrium outcome.

All the assumptions previously defined apply to this part. Moreover, the relationship between the error term variances has to de described. For this purpose, remembering that patients' choice is driven by the perceived quality $\tilde{q}$ which is biased because of the error ε_i, the product of the variances of the error term $\left(\sigma_a^2 \cdot \sigma_b^2 = 1\right)$ is assumed constant and equal to 1 but the ratio $\frac{\sigma_a^2}{\sigma_b^2}$ of variances is allowed to vary. This quoted ratio value is reported in the second column of Table 4. Remember that the error term affecting the perceived quality is characterized by heteroskedasticity, i.e., its variance varies according to the distance from the provider of medical care.

In Table 4 different scenarios of cost and variance are matched up. In the first column, as provided in previous tables, one can read the marginal cost relationship between hospitals. The second column considers different variance scenarios characterized by different variances in the error term of the equation describing the perceived quality (see Sect. 5). In particular, when the ratio c_a/c_b is equal to 1, the hospitals face the same marginal cost, or, in other words, they show the same level of efficiency. It is possible to note the negative consequences of quality observation bias with respect to the benchmark provided in Table 1. In fact the quality level is lower (if compared with the equilibrium values of Table 1) because of the term $-\frac{\beta}{3\alpha}\left(\sigma_i^2 + 2\sigma_j^2\right)$ (see Eq. 17) which enters the utility function and represents patients uncertainty on quality. Obviously this turns out to be a gain for the hospitals in terms of larger profits and a loss for patients in terms of a reduction in their surplus. In fact, the quality perception bias reduces the hospital incentive to compete via quality for patients. This comes from the fact that, by setting the heteroskedasticity in the error term variance, the result is, in practice, a greater disutility in distance. The latter increases the location rent for hospitals that find it profitable to skimp on quality.

The scenarios of Table 4 where $\frac{\sigma_a^2}{\sigma_b^2} = 1$ are the only ones comparable with those provided in previous sections. In fact, the other rows of Table 4 are characterized (differently from Tables 1, 2, 3) by a different ratio of the error term variances (that are not detectable in the case of perfect information).

Looking at Table 4 it is possible to note that an increase in the competitor's variance (with respect to its own variance) allows a gain in terms of profit for the hospitals with the lower variance. On the other hand as long as we move from the condition of equality in variances we record a patients' loss in terms of utility. However, from the social welfare perspective, the latter is compensated by the profit gain of hospitals. From the hospital perspective a large variance in the perceived quality is an opportunity because it allows for greater profits, however a larger variance with respect to the competitor's is a cost. In fact, the hospital with the larger variance is forced to increase its quality level and as a consequence its costs will increase. On the other hand, when the variance is low, it is possible to serve a larger share of the market even if providing a lower quality.

Thus hospitals benefit from the market uncertainty but in order to maximize their profit they wish to reduce (possibly below the competitor's level) their own variance (in the perceived quality they offer) so as to take the maximum advantage from this kind of asymmetry in information. By this effort/strategy they are able to

Table 4 Simulation 4

$\frac{c_a}{c_b}$	$\frac{\sigma_a^2}{\sigma_b^2}$	q_a	q_b	$\sum_{i=a}^{b} q_i$	π_a	π_b	$\sum_{i=a}^{b} \pi_i$	D_a	D_b	S_a	S_b	$\sum_{i=a}^{b} S_i$	W
0.8	0.7	1.22	1.11	2.33	3.75	2.60	6.34	0.57	0.43	0.40	0.27	0.673	7.02
	0.8	1.26	1.10	2.35	3.66	2.65	6.31	0.57	0.43	0.41	0.28	0.683	6.99
	0.9	1.28	1.08	2.36	3.59	2.70	6.29	0.56	0.44	0.41	0.28	0.688	6.98
	1	1.30	1.06	2.36	3.53	2.75	6.29	0.56	0.44	0.41	0.27	0.689	6.97
	1.1	1.31	1.05	2.36	3.49	2.80	6.29	0.55	0.45	0.41	0.27	0.687	6.98
	1.2	1.33	1.03	2.36	3.44	2.85	6.30	0.55	0.45	0.41	0.27	0.682	6.98
	1.3	1.34	1.01	2.35	3.41	2.90	6.31	0.55	0.45	0.41	0.27	0.676	6.99
0.9	0.7	1.15	1.15	2.30	3.56	2.81	6.37	0.54	0.46	0.35	0.30	0.650	7.02
	0.8	1.18	1.14	2.32	3.47	2.87	6.34	0.54	0.46	0.36	0.30	0.662	7.00
	0.9	1.20	1.13	2.33	3.40	2.92	6.32	0.53	0.47	0.36	0.30	0.667	6.99
	1	1.22	1.11	2.33	3.34	2.97	6.32	0.53	0.47	0.37	0.30	0.668	6.98
	1.1	1.24	1.09	2.33	3.29	3.03	6.32	0.52	0.48	0.37	0.30	0.666	6.99
	1.2	1.25	1.08	2.33	3.25	3.08	6.33	0.52	0.48	0.37	0.30	0.662	6.99
	1.3	1.26	1.06	2.32	3.22	3.13	6.35	0.52	0.48	0.36	0.17	0.656	7.00
1	0.7	1.09	1.21	2.29	3.38	3.00	6.38	0.51	0.49	0.32	0.33	0.643	7.02
	0.8	1.12	1.19	2.31	3.29	3.06	6.35	0.51	0.49	0.32	0.33	0.655	7.00
	0.9	1.14	1.18	2.32	3.22	3.11	6.33	0.50	0.50	0.33	0.33	0.661	6.99
	1	1.16	1.16	2.32	3.16	3.16	6.32	0.50	0.50	0.33	0.33	0.662	6.99
	1.1	1.18	1.15	2.32	3.11	3.21	6.33	0.50	0.50	0.33	0.33	0.661	6.99
	1.2	1.19	1.13	2.32	3.07	3.27	6.34	0.49	0.51	0.33	0.33	0.657	7.00
	1.3	1.20	1.11	2.31	3.04	3.32	6.35	0.49	0.51	0.33	0.32	0.652	7.01
1.1	0.7	1.04	1.26	2.30	3.21	3.16	6.37	0.49	0.51	0.29	0.36	0.646	7.02
	0.8	1.07	1.25	2.32	3.12	3.22	6.34	0.48	0.52	0.30	0.36	0.659	7.00
	0.9	1.10	1.23	2.33	3.05	3.27	6.32	0.48	0.52	0.30	0.36	0.665	6.99
	1	1.12	1.22	2.33	2.99	3.33	6.32	0.47	0.53	0.30	0.36	0.667	6.98
	1.1	1.13	1.20	2.33	2.94	3.38	6.32	0.47	0.53	0.31	0.36	0.666	6.99
	1.2	1.14	1.18	2.32	2.90	3.43	6.33	0.47	0.53	0.30	0.36	0.663	7.00
	1.3	1.15	1.16	2.31	2.87	3.48	6.35	0.46	0.54	0.30	0.35	0.658	7.01
1.2	0.7	1.00	1.32	2.32	3.05	3.31	6.35	0.47	0.53	0.27	0.39	0.658	7.01
	0.8	1.04	1.30	2.34	2.96	3.36	6.32	0.46	0.54	0.28	0.39	0.671	6.99
	0.9	1.06	1.29	2.35	2.89	3.42	6.30	0.46	0.54	0.28	0.40	0.678	6.98
	1	1.08	1.27	2.35	2.83	3.47	6.30	0.45	0.55	0.28	0.40	0.680	6.98
	1.1	1.09	1.25	2.35	2.78	3.52	6.30	0.45	0.55	0.28	0.39	0.679	6.98
	1.2	1.11	1.24	2.34	2.74	3.57	6.31	0.44	0.56	0.28	0.39	0.676	6.99
	1.3	1.11	1.22	2.33	2.71	3.62	6.33	0.44	0.56	0.28	0.39	0.672	7.00

contain costs by reducing the quality level they have to provide in order to maximize their profits and serve a larger portion of the market.

It is also clear that the fact that, because we have heteroskedasticity and that the variance of the error term is affected by distance, a more effective monopolistic rent (because of the spatial differentiation of the market) is detectable. This is the consequence of the fact that the quality level provided is noticeably lower with respect to the other scenarios. In this setting the quality "attractive power" is considerably reduced. To this extent, it is not convenient for hospitals to invest too much on (costly) quality but it would be better to invest on (cheaper) advertising, exploiting information asymmetry to their own advantage (Montefiori 2008). Hospitals have an interest in controlling variance and reducing asymmetric information. Reducing their quality variance would determine a direct increase in profit. Thus they will invest money in "information activity". We can also observe that both the hospitals and the purchaser have an interest in reducing information asymmetry. The hospital aims to reduce the variance in its own quality in order to boost its demand (even when low quality levels are provided), the purchaser aims to decrease uncertainty to avoid hospital incentives on quality curbing.

6.5 Uncertainty: Asymmetric Objectives and Asymmetric Costs

In this section the conditions of asymmetric objectives and asymmetric costs (and their interactions) in a context of uncertainty are investigated by numerical simulation. The results are provided in Table 5.

In this scenario uncertainty only affects the quality provided by the private hospital B, by a lower optimal level set for quality. On the contrary, the behavior of public hospital A, intended at quality maximization (the "excellence" goal), is not affected by uncertainty.

Uncertainty allows for higher profits for the private hospital even if the behavior of the public hospital considerably reduces its scope. If we compare the case in which $c_a/c_b = 1$ and $\frac{\sigma_a^2}{\sigma_b^2} = 1$ of Table 5 with those of Table 4 we can

Table 5 Simulation 5

$\frac{c_a}{c_b}$	$\frac{\sigma_a^2}{\sigma_b^2}$	q_a	q_b	$\sum_{i=a}^{b} q_i$	π_a	π_b	$\sum_{i=a}^{b} \pi_i$	D_a	D_b	S_a	S_b	$\sum_{i=a}^{b} S_i$	W
0.8	0.7	3.54	2.26	5.80	0	0.28	0.28	0.86	0.14	2.36	0.30	2.657	2.94
	1	3.54	2.18	5.72	0	0.37	0.37	0.84	0.16	2.26	0.33	2.588	2.96
	1.3	3.54	2.11	5.65	0	0.45	0.45	0.82	0.18	2.18	0.35	2.529	2.98
0.9	0.7	3.33	2.25	5.58	0	0.47	0.47	0.81	0.19	2.10	0.38	2.485	2.95
	1	3.33	2.17	5.50	0	0.58	0.58	0.79	0.21	2.01	0.41	2.420	3.00
	1.3	3.33	2.10	5.43	0	0.68	0.68	0.78	0.22	1.94	0.42	2.365	3.04
1	0.7	3.16	2.24	5.41	0	0.66	0.66	0.77	0.23	1.89	0.45	2.349	3.01
	1	3.16	2.16	5.32	0	0.79	0.79	0.75	0.25	1.81	0.48	2.287	3.08
	1.3	3.16	2.09	5.25	0	0.90	0.90	0.73	0.27	1.74	0.49	2.235	3.14
1.1	0.7	3.02	2.25	5.26	0	0.85	0.85	0.73	0.27	1.72	0.52	2.239	3.09
	1	3.02	2.17	5.18	0	1.00	1.00	0.71	0.29	1.64	0.54	2.181	3.18
	1.3	3.02	2.10	5.11	0	1.12	1.12	0.70	0.30	1.58	0.55	2.131	3.25
1.2	0.7	2.89	2.26	5.14	0	1.04	1.04	0.70	0.30	1.57	0.58	2.149	3.19
	1	2.89	2.18	5.06	0	1.20	1.20	0.68	0.32	1.50	0.60	2.094	3.29
	1.3	2.89	2.11	4.99	0	1.33	1.33	0.66	0.34	1.44	0.60	2.046	3.37

observe a striking difference in hospital B profits. The behavior of hospital A which aims at excellence noticeably reduces the location rent of hospital B which is forced to keep up the quality level provided in order to retain market share. In other words, it is possible to state that the presence of a publicly owned hospital aiming at quality maximization in a health care market characterized by uncertainty is able to contain the negative effects (in terms both of quality and patient surplus) determined by information asymmetry.

7 Conclusions

This work has shown by a theoretical analysis and numerical simulations the role played by heterogeneous objectives, asymmetric costs and uncertainty in affecting the equilibrium outcome in a duopoly market for health care in which hospitals compete à la Hotelling for patients.

The analysis implemented draws from the ascertainment that globalization is affecting the market for health care. Very soon, hospitals that are very different in terms of ownership/objectives and costs (such as public hospitals, for profit and non-profit private hospitals, teaching hospitals) will compete in a spatially differentiated market.

However very little is known about the consequences of this competition and hospitals (as well as Countries) still seem to be unprepared to cope with this new framework.

The analysis provided in the paper aims to put some light on this new scenario. In particular, the understanding of patient and hospital behavior, with particular reference to patient disutility on distance and quality observation bias, are aspects of great relevance which deserve the attention of the policy maker. The aim is to provide, through the analysis of the present work, new informative tools at the policy maker's disposal, to implement an efficient and effective regulation activity.

For this reason the aspects on which the paper has mainly focused are patient mobility (rigidity) and uncertainty. With reference to the latter it has to be noticed that the literature refers to health care as a "*credence good*" (Montefiori 2008), i.e., a good whose quality cannot be correctly evaluated by consumers/patients even after they have experienced the services. The role played by uncertainty in affecting the market for care is therefore evident, a fortiori when medical care is provided by hospitals located far away from the patient who has to judge them.

The study provides the result that asymmetric costs, in the case that both hospitals are profit maximizers, allow for a social welfare improvement. The presence of a more efficient hospital, i.e., a hospital characterized by lower values of marginal cost, sets the conditions for an overall quality improvement. In fact, if on the one hand, the more efficient hospital gain shares of the market by increasing the quality it provides, on the other the less efficient hospital is forced to keep quality upwards in order to reduce its loss (via the demand mechanism which is quality driven). This result is amplified when the patient mobility is high but it is limited in scope when the patient mobility is low. Note that patient mobility depends on two key elements: patients' utility on quality and patients disutility on distance.

The chapter shows that the presence of a publicly managed hospital which pursues the objective of quality maximization (*excellence*) is able to ensure, in general, higher levels of quality, patient surplus and welfare. This result comes from the fact that the private hospital is forced to react to the public hospital quality setting by investing on costly quality and, in so doing, avoiding a large loss of market share. However, in the limiting case of a very high level of public hospital inefficiency, the welfare gain may change in sign and the undesirable effect could even be an equilibrium outcome less desirable than the benchmark (in which only profit maximizers hospitals are in).

Another aspect of relevance treated in the article concerns the role of uncertainty in affecting the equilibrium. Because it is unrealistic to assume that patients have the ability to perfectly observe the quality provided by hospitals, the concept of *perceived quality* is introduced. Specifically, a particular type of error characterized by a heteroskedastic variance, has been used to investigate hospital behavior and market equilibrium.

What the study highlights are the negative consequences belonging to the presence of asymmetric information in the health market. In particular, the presence of uncertainty reduces the overall quality, decreases the patient surplus and increases the hospital's profit. Hospitals tend to take advantage from uncertainty curbing their costly quality and, in doing so, increasing their profit. Uncertainty causes an increase in the rigidity of patient mobility, benefitting the hospitals with a more effective monopolistic rent as a direct consequence of the market differentiation.

However the presence of publicly owned (and managed) hospitals aiming at quality maximization (the reference is to the case of excellence previously quoted) is able to contrast the location rent strengthened by uncertainty, and, at the same time, contain the negative effects (in terms of quality and consumers' surplus) caused by the asymmetry of information.

References

Appleby, J., & Dixon, J. (2004). Patient choice in the NHS. *British Medical Journal, 329*, 61–62.

Beato, P., & Mas-Colell, A. (1984). The marginal cost pricing rule as a regulation mechanism in mixed markets. In M. Marchand, P. Pestieau, & H. Tulkens (Ed.), *The performance of public enterprises: Concepts and measurements*. Amsterdam: Kluwer Academic Publishers.

Brekke, K., Cellini, R., Siciliani, L., & amd Straume, O. R. (2012a). Competition in regulated markets with sluggish beliefs about quality. *Journal of Economics and Management Strategy, 21*(1), 131–178.

Brekke, K., Levaggi, R., Siciliani, L., & Straume, O.R. (2011a). Patient mobility, health care quality and welfare, NHH discussion paper n.13.

Brekke, K., Nuscheler, R., & Straume, O. R. (2006). Quality and location choices under price regulation. *Journal of Economics and Management Strategy, 15*, 207–227.
Brekke, K. R., Siciliani, L., & Straume, O. R. (2011b). Hospital competition and quality with regulated prices. *Scandinavian Journal of Economics, 113*, 444–469.
Brekke, K. R., Siciliani, L., & Straume, O. R. (2012b). Quality competition with profit constraints. *Journal of Economic Behavior & Organization, 84*, 642–659.
Chalkley, M., & Malcomson, J. M. (1998a). Contracting for health services with unmonitored quality. *The Economic Journal, 108*, 1093–1110.
Chalkley, M., & Malcomson, J. M. (1998b). Contracting for health services when patient demand does not reflect quality. *Journal of Health Economics, 17*, 1–19.
Gravelle, H. (1999). Capitation contracts: access and quality. *Journal of Health Economics, 18*, 315–340.
Gravelle, H., & Masiero, G. (2000). Quality incentives with imperfect information and switching costs in a regulated market: capitation in general practice. *Journal of Health Economics, 19*, 1067–1088.
Gravelle, H., & Masiero, G. (2002). Quality incentives under a capitation regime. The role of patient expectations. In L. Bjorn (Ed.), *Individual decisions for health*. London: Routledge.
Gravelle, H., & Sivey, P. (2010). Imperfect information in a quality-competitive hospital market. *Journal of Health Economics, 29*, 524–535.
Herr, A. (2011). Quality and welfare in a mixed duopoly with regulated prices: the case of a public and a private hospital. *German Economic Review, 12*, 422–437.
Hotelling, H. (1929). Stability in competition. *The Economic Journal, 39*, 41–57.
Levaggi, R. (2005). Hospital health care: cost reimbursement and quality control in a spatial model with asymmetry of information. *International Journal of Health Care Finance and Economics, 5*(4), 327–349.
Levaggi, R. (2007). Regulating internal markets for hospital care. *Journal of Regulatory Economics, 2*(2), 173–193.
Levaggi, R., & Montefiori, M. (2013). Patient selection in a mixed oligopoly market for health care: The role of the soft budget constraint. *International Review of Economics, 60*(1), 49–70.
Ma, C. A. (1994). Health care payment systems: Cost and quality incentives. *Journal of Economic Management and Strategy, 3*, 93–112.
Montefiori, M. (2005). Spatial competition for quality in the market for hospital care. *European Journal of Health Economics, 6*(2), 131–135.
Montefiori, M. (2008). Information versus advertising in the market for hospital care. *International Journal of Health Care Finance and Economics, 8*(3), 145–162.
Montefiori, M., & Resta, M. (2009). A computational approach for the health care market. *Health Care Management Science, 12*(4), 344–350.
Palm, W., & Glinos, I. A. (2010). Enabling patient mobility in the EU: Between free movement and coordination, In E. Mossialos, G. Permanand, R. Baeten, & T. K. Hervey (Eds.), *Health systems governance in Europe, the role of European Union law and policy*. Cambridge: Cambridge University Press.
Sanjo, Y. (2009). Quality choice in a health care market: A mixed duopoly approach. *European Journal of Health Economics, 10*, 207–215.
Vrangbaek, K., & Ostergren, K. (2006). Patient empowerment and the introduction of hospital choice in Denmark and Norway. *Health Economics, Policy and Law, 1*(4), 371–394.

Cross Border Health Care Provision: Who Gains, Who Loses

Rosella Levaggi and Francesco Menoncin

Abstract The diffusion of the welfare state has produced a widespread involvement of the public sector in financing the production of private goods for paternalistic reasons. In this chapter we model the production of health care as a merit impure local public good whose consumption is subsidized and whose access is free, but not unlimited. The impure local public good aspect means that the production of health care spreads its benefits beyond the geographical boundaries of the Region where it is produced. Finally, we include the (optional) provision of an equalization grant that allows reduction of fiscal imbalance among Regions. In this framework we study the possible effects of cross border provision of health care. We assume that information is complete and symmetric and that there is no comparative advantage in local provision. In this context devolution is always suboptimal for the whole community: the lack of coordination means that the impure public good is under-provided. However, more efficient Regions may be better off because of the impure public good nature of health care.

Keywords Health care provision · Decentralization · Impure public goods · Cross border supply

JEL Classification H77 · H4

R. Levaggi (✉) · F. Menoncin
Department of Economics and Management, University of Brescia, Via S. Faustino 74/b
25122 Brescia, Italy
e-mail: levaggi@eco.unibs.it

F. Menoncin
e-mail: menoncin@eco.unibs.it

R. Levaggi and M. Montefiori (eds.), *Health Care Provision and Patient Mobility*, Developments in Health Economics and Public Policy 12,
DOI: 10.1007/978-88-470-5480-6_10,

1 Introduction

The diffusion of the welfare state has produced a widespread involvement of the public sector in financing the production of private goods for paternalistic reasons (Schnellenbach 2012). Because of the parallel process of devolution, the provision of such goods has often been delegated to local Governments and provided mainly through (partial or total) public subsidies and expenditure-based equalization grants.

Health care is one of the most relevant examples in this class of goods; its expenditure has been steadily growing since the inception of the Welfare State and it is expected to grow in the future. Whatever the cause for such a growth, the recession that started in 2008 calls for all possible efforts to reduce its cost. Telemedicine and the widespread diffusion of internet technology may allow substantial reduction of diagnostic costs. In the US, where the cost of health care has reached about 18 % of GDP, a growing number of insurance companies and Health Maintenance Organisations (HMOs) hire physicians in developing countries and use them to make diagnoses.

Smith et al. (2009) present the four models of service delivery involving cross country trade in health care which are regulated by GATS (General Agreement on Trade and Service).

The first model is the supply of cross border health services, an emerging trade that has been made possible by the advance in information technology. It consists in remotely providing a service from a provider in one country to an overseas recipient. It covers a widening basket of services ranging from diagnostics (teleradiology and laboratory testing) to treatment (remote surgery and teleconsultation). The second model is about consumption of services abroad and it comprises what is normally meant by "patient mobility" in all the forms that have been presented in this book. The third model can be defined as "foreign direct investment" and is somehow related to the second model. In order to deliver services with a standard level of quality, new hospitals devoted to treating non-residents are built in developing countries by joint ventures between domestic and foreign partners. Finally, the fourth model involves health professionals that move from one country to another. The first model is becoming popular in the US where the cost of health care is particularly high. In order to reduce the insurance bills to their clients, some companies have hired specialists in emerging countries. The tests are carried out in the US by nurses who send the images abroad. In the emerging country doctors examine them and send the referral back to the US. In this way the cost of the medical staff is sensibly reduced without lowering the quality. In Europe the second and the fourth model of trade described above are the most common. In this chapter we concentrate on the first two models and we build a framework that allows us to study cross border health services in a context where equity issues are also taken into account. In this chapter we model the production of health care as a merit impure local public good whose consumption is subsidised and whose access is free, but not unlimited. The impure local public good aspect means that the production of health care spreads its benefits beyond the geographical boundaries of the Region where it is produced. Finally, we include the (optional) provision of an equalization grant that allows reduction of fiscal imbalance among Regions. In this framework we study the possible effects of cross border provision of health care.

Our benchmark model is *decentralization*, where provision at local level is made by each Region separately and where cross border mobility is not allowed; we will then study the welfare implications of allowing mobility and some form of coordination in the expenditure decisions among the Regions. We assume that information is complete and symmetric and that there is no comparative advantage in local provision.[1] In this context devolution is always sub-optimal for the whole community: the lack of coordination means that the impure public good is under-provided. Opening to international trade allows the less efficient Region to improve its welfare, but the most efficient one may experience a welfare loss due to the fact that the price for cross border shopping may be lower than the production cost. This effect is due to the impure public good characteristic of health care. On the other hand, international trade and more coordination among Regions (either through bilateral agreements or the intervention of an upper Government level) will ensure that spillovers are taken into account. In general, health care expenditure will increase and so will the equalization grant. The richer Region may suffer a welfare loss in this case because more resources will be required to finance the grant to the poorer Region. This opens an interesting policy debate about cross border supply and federalism. Unless this policy is imposed by an upper level, it is not likely to be the outcome of coordination between local authorities. This may partially explain the present debate at EU level about patient mobility across countries (Legido-Quigley et al. 2007, 2012; Glinos et al. 2010; Brekke et al. 2011).

This chapter is organised as follows: in Sect. 2 we present the general framework. In Sect. 3 the first best centralized solution is computed, while Sect. 4 presents devolution. A discussion and a numerical simulation are presented in Sect. 5. Section 6 concludes.

2 The Model

We model health care as an impure local public good with spillovers. Impure public goods belong to a fairly heterogeneous category, varying from impure public goods in their most traditional definition (Musgrave and Musgrave 1989) to

[1] The traditional literature assumes that there is a comparative advantage in producing at local level. Asymmetry of information and spillovers may however reduce this advantage (see Koethenbuerger 2008; Tommasi and Weinschelbaum 2007).

paternalistic goods with spillovers.[2] In our model, this characteristics of the service is captured by the form of the subsidy. For an impure public good, the usual form is a user charge, i.e. the consumer is asked to pay a fraction of the price of the service produced. When the impure public good is also a paternalistic good, it is usually supplied free of charge, but not necessarily to the entire population; this is the approach we will be using in this work. Let us consider a community, whose population is normalized to one and equally divided into two Regions $i \in \{1,2\}$ (which can be two Local Authorities in the same country or two Member States in a Union). Each individual has an exogenous money income, M^k in the range $[\underline{M}_i, \overline{M}_i]$, with density function $\varphi_i(M^k)$. Then, total income in local authority i is:

$$Y_i = \frac{1}{2}\int_{\underline{M}_i}^{\overline{M}_i} M^k \varphi_i(M^k) dM^k.$$

Income it is used to buy private commodities and one or zero unit of a paternalistic local public good (health care). Each inhabitant has a preference towards such a good, which is defined by the parameter $\alpha \in [0, \beta]$ which also measures the utility of each unit consumed. We assume α is uniformly distributed between agents (i.e. its density function is β^{-1}).

Health care is a paternalistic good whose access is restricted to individuals with $\alpha > p_i$, where p_i represents the marginal utility of the paternalistic good the decision-maker is willing to finance. It is financed using a linear income tax at rate τ_i. The utility function for a representative individual living in local authority i can be written as:

$$V_i^k = M^k(1-\tau_i) + \begin{cases} \alpha^k & \text{if } \alpha^k \geq p_i, \\ 0 & \text{if } \alpha^k < p_i. \end{cases} + \phi_i(S_i, S_j). \tag{1}$$

Finally, utility from the public good nature of health care depends on the quantity produced in each Region (S_i), which accrues welfare because it creates a local public good in terms of option to use health care if needed.[3] We assume that preferences for the impure public good have the following form:

$$\phi_i(S_i, S_j) = w_i(z_i S_i - S_i^2) + (1-w_i)\left(z_i S_j - S_j^2\right),$$

where S_i and S_j are the quantity of good H produced in the two jurisdictions.

The function is increasing and concave in its arguments (decreasing marginal utility at community level), hence the utility of an additional unit depends on

---

[2] The basic difference between a merit good and an impure public good is that the former is actually a private good that is used to improve income redistribution or to pivot consumers' preferences towards the use of goods which the planner thinks they should use. We define as spurious merit good a class of services that have this *dual* characteristics, for example health, education and cultural activities.

[3] The literature had defined this an "option demand good", see Lindsay (1969).

where it is produced.[4] The level of publicness of the good depends on the value of parameter w_i. In particular:

1. for $w_i = \frac{1}{2}$, the good is a public good;
2. for $w_i = 1$, the good is a local public good;
3. for $0 < w_i < 1_i$, the good is a local public good with spillovers.

z determines the marginal utility of health care produced in each Region. The marginal utility is in fact equal to $z - 2S$ and z has to be sufficiently high to insure that $z - 2S > 0$.

The quantity of health care demanded for in each Region is

$$Q_i = \frac{1}{2}\int_{p_i}^{\beta} \frac{1}{\beta} d\alpha = \frac{\beta - p_i}{2\beta},$$

while the utility given by consuming Q_i is

$$\frac{1}{2}\int_{p_i}^{\beta} \frac{1}{\beta}\alpha d\alpha = \frac{\beta^2 - p_i^2}{4\beta}.$$

Total welfare of Region i is given by the sum of the net income, the utility from consumption of the locally produced good, and the utility from the public characteristics of the locally produced good:

$$W_i = Y_i(1 - \tau_i) + \frac{\beta^2 - p_i^2}{4\beta} + w_i\left(z_i S_i - S_i^2\right) + (1 - w_i)\left(z_i S_j - S_j^2\right). \quad (2)$$

We assume:

1. $Y_1 > Y_2$, i.e. Region 1 is richer than 1;
2. the marginal cost to locally produce the good is constant and there is no fixed cost, but also in this case Region 1 is more efficient than Region 2.

Given the double nature of private and public good, the beneficiaries of the two characteristics may not coincide. The quantity demanded by residents in each local authority (Q_i) does not necessarily need to coincide with the quantity produced in the same area (S_i). In other words, we allow for cross border provision.[5] When cross border mobility is allowed, supply and demand are matched by Regions through negotiation of a transfer price which may not be equal to marginal cost, given the externality produced.

---

[4] For a distinction between global public goods and local public goods with spillovers, see Levaggi (2010).

[5] The analysis presented here neglects transportation costs in order to concentrate on the coordination problem among regions.

2.1 Equality Issues

In most countries income is unevenly distributed across geographical areas and the super-national authority may be called to introduce some form of interregional distribution among Regions. This function is certainly more important within a unitary State, but in Europe also the EU plays an important role in this process.[6] In our model we introduce horizontal equity which implies that a given tax effort should be rewarded with the provision of a uniform amount of public services. This objective can be pursued using an equalization grant G_i. Several forms of equalization exist; in this model we will use the lump-sum form as suggested by Dahlby and Wilson (1994) and Smart (1998). The idea behind this equalization grant is to virtually increase the tax base of the poorer Region so that the fiscal effort in terms of tax rate allows the same tax revenue to be obtained. The lump sum form is chosen in order to reduce the deadweight loss deriving from the intergovernmental grant, but this implies that the amount of the equalization grant depends on the total level of health care expenditure. The equalization grant G_i can be written as :

$$G_i = \frac{1}{2}\tau^m\left(\overline{Y} - Y_i\right),$$

where

$$\tau^m = \frac{\tau_1 Y_1 + \tau_2 Y_2}{Y_1 + Y_2},$$
$$\overline{Y} = \frac{Y_1 + Y_2}{2},$$

are the national average tax rate and the standardized tax base respectively. Both are invariant to each regional fiscal decision, i.e. local authorities do not perceive the effects that their tax rate has on the equalization grant. It is interesting to note that this form of equalization grant implies that $G_i = -G_j$. It is a form of horizontal equalization grant since it does not imply raising a tax at the super national level. This form is particularly suitable to study welfare in the EU where taxes are levied at the lower level.

Given the local tax τ_i and the grant G_i, if we assume that the total cost of producing the quantity Q_i is $v_i Q_i$, then the local authority constraint can be written as:

$$\tau_i = \frac{v_i Q_i - G_i}{Y_i}.$$

[6] About 30 % of the total EU budget is in fact devoted to Regional support (De la Fuente et al. 2010).

3 Decentralized Solution

Each Region sets its own tax rate and service production according to its preferences and resources and takes G_i as given. The upper government level sets the equalization grant; this actor is the last one to move, i.e. it sets the grant after the Regions have set their own level of expenditure and taxation.

This solution is not optimal from a welfare point of view because the Regions do not fully take into account the consequences of their actions on welfare (Petretto 2000). This is a well known result in the literature on fiscal federalism: the presence of spillovers means that the quantity produced in each local authority is always lower than the optimal one. Below we examine the two cases with and without cross border supply and we will then consider the implications for welfare of opening health care to trade.

3.1 Without Cross Border Supply

Let us now examine the optimal conditions in an environment where each Region maximizes its own welfare and cross border supply is not allowed. The problem can be written as:

$$\begin{gathered} \max_{p_i} W_i(p_i, Q_i, Q_j) \\ \text{s.t.} \\ \tau_i = \frac{v_i Q_i - G_i}{Y_i}, \\ Q_i = \frac{\beta - p_i}{2\beta}. \end{gathered} \tag{3}$$

The optimal provision for the problem are derived in Appendix and can be written as

$$\begin{aligned} \bar{p}_i &= \beta \frac{v_i + w_i(1 - z_i)}{\beta + w_i} = v_i - w_i\left(\frac{v_i + \beta(1 - z_i)}{\beta + w_i}\right), \\ \bar{Q}_i &= \frac{1}{2} - \frac{v_i - w_i(1 - z_i)}{2(\beta + w_i)}. \end{aligned} \tag{4}$$

The number of people that receive health care depends on the marginal cost of production (v_i) and the public good aspect of health care w_i. This implies that the cost effectiveness of the treatment has to include both the utility accruing to the patient and the increased benefit that society receives from the production of that specific unit of health care.

The quantity produced is not optimal because each Region does not take into account the positive spillover its production is creating on the neighbour Region and it does not exploit the higher productivity level of its neighbour.

Total welfare in this case can be written as:

$$\begin{aligned}\overline{W}_i &= Y_i(1-\bar{\tau}_i) + \frac{\beta^2 - p_i^2}{4\beta} + w_i\left(z_i\bar{Q}_i - \bar{Q}_i^2\right) + (1-w_i)\left(z_i\bar{Q}_j - \bar{Q}_j^2\right),\\ \bar{W} &= \sum_{i=1,2} \overline{W}_i,\\ \bar{\tau}_i &= \frac{\bar{Q}_i v_i}{Y_i} - \frac{\left(\overline{Y} - Y_i\right)}{(Y_1+Y_2)Y_i}\sum_{i=1,2}\bar{Q}_i v_i.\end{aligned} \tag{5}$$

This framework can be considered our benchmark model. In the following sections we will study the effects of cross border supply of health care on this solution in terms of quantity of health care supplied, its regional distribution and its implications for welfare. Cross border mobility may result either from bilateral agreements between the Regions or from specific upper Government tier initiatives. In our model we do not distinguish between these two sources; in fact we concentrate on the implications of the decision process rather than on how it was originated.

In a neoclassical model, trade is always beneficial. Because of specialization, both trade partners will be better off (see, for instance, Frankel and Romer 1999). In this model this might not be the case owing to the combined effect of several factors: the impure public good characteristics of health care and the presence of the equalization grant.

3.2 *With Cross Border Supply*

Transnational health care is a new and emerging phenomenon in the European context, but it may assume several forms that needs to be studied separately. Levaggi and Levaggi (2014) show the different welfare implications of regulated patient mobility versus patient choice. In this paper we focus on a more specific form of patient choice by considering forms of collaboration between nations very close to one another or the flow of mobility across neighbouring regions within the same country. In both contexts we assume that the traveling costs are negligible and can be approximated to zero. In this way, we can concentrate on the welfare effects of opening the borders to competition. Usually, increased competition should improve welfare, but this may not be the case in a context where the good supplied is subsidized. In our model two forms of redistribution exist:

- interpersonal redistribution: the good is supplied for free to those that need health care and it is financed out of general taxation;
- inter-regional redistribution: expenditure is used to equalize resources across Regions.

In this context, total welfare will certainly increase (due to a more efficient use of resources), but the benefits between the two Regions may be distributed quite unevenly and opening to international trade may not be a Pareto optimal solution. In this section we study such a case.

When the service is supplied across borders, the Region producing it is reimbursed at rate q (the so-called "transfer price"). Accordingly, the local authority i receives: (i) the amount of tax $Y_i\tau_i$, (ii) the grant G_i and, on the other side, must pay: (i) the production cost v_iS_i, and (ii) the reimbursement q proportional to the product supplied across the border $Q_i - S_i$. The constraint can be written as

$$Y_i\tau_i + G_i = v_iS_i + q(Q_i - S_i),$$

from which we have

$$\tau_i = \frac{v_iS_i + q(Q_i - S_i) - G_i}{Y_i}.$$

The problem faced by each local authority can thus be written as:

$$\begin{aligned} &\max_{p_i,S_i} W_i(p_i, S_i, S_j) \\ &\text{s.t.} \\ &\tau_i = \frac{v_iS_i + q(Q_i - S_i) - G_i}{Y_i}. \end{aligned} \tag{6}$$

The FOCs for the problem are derived in Appendix and can be written as

$$\begin{aligned} p_i &= q, \\ S_i &= \frac{q - v_i + z_iw_i}{2w_i}. \end{aligned} \tag{7}$$

The local decision-maker does not take into account the spillover effect that its production creates on the neighbouring jurisdiction. Furthermore, in their maximization process, they take q as a given parameter, but in equilibrium only one value exists which clears the market. To reconcile devolution with market clearing conditions, it is necessary to find the q that satisfies the optimal conditions (7) and the market clearing constraint. The problem can be solved using a Nash game:

$$\begin{aligned} S_i &= \frac{q - v_i + z_iw_i}{2w_i}, \\ S_1 + S_2 &= Q_1 + Q_2 = 1 - \frac{q}{\beta}, \end{aligned} \tag{8}$$

and the level of transfer price that clears the market is equal to:

$$\begin{aligned} p_i &= q, \\ S_i &= \frac{q - v_i}{2w_i} + \frac{z_i}{2}, \\ q &= \frac{w_2(\beta + w_1)\bar{p}_1 + w_1(\beta + w_2)\bar{p}_2}{w_2(\beta + w_1) + w_1(\beta + w_2)}, \end{aligned} \tag{9}$$

where $\bar{p}_i = \beta \frac{v_i + w_i(1-z_i)}{\beta + w_i}$, $i = 1, 2$ is the utility of the marginal patients receiving care in the decentralized solution. The first interesting result of this solution is that the level of production in the two Regions is now determined by the transfer price q. From Eq. (9) we can observe that the transfer price is a weighted average of $\bar{p}_i$. As in the previous model, the number of people receiving care depends on the marginal price of production and on the public good aspect of health care. This imply that the number of people that receive health care in the new equilibrium will increase in Region 2 and will decrease in Region 1. Overall the number of patients treated is however increasing because the average cost to treat them is decreasing. To determine the effects on demand of cross border mobility, the difference between Eqs. (4) and (9) should be evaluated. The latter depends on the combined effect of the difference in price and in the preferences. However, some conclusions can be drawn[7]:

- if the good has the same local public good characteristic ($w_i = w_j = w$), then $p_i > q$ if $z_j < z_i + \frac{v_j - v_i}{w}$;
- if the preferences for the public good are uniform ($z_i = z_j$), then $p_i > q$ if $v_i - v_j < 0$;
- in any case $q < \min(v_i, v_j)$ which means that health care is always subsidized.

The last point is quite interesting because it means that in a context of cross border mobility the Region that receives patients will have to finance its costs with local resources, an element that depends on the impure public good nature of health care, and that has an important role in the distribution of the benefits arising from cross border shopping.

After finding $\hat{q}$ that clears the market, it will be possible to obtain $\widehat{p}$, $\widehat{S_1}$ and $\widehat{S_2}$. Total welfare can be written as:

$$\begin{aligned} \hat{W} &= Y_i(1-\hat{\tau}_i) + \frac{\beta^2 - \hat{p}_i^2}{4\beta} + w_i\left(z_i\hat{S}_i - \hat{S}_i^2\right) + (1-w_i)\left(z_i\hat{S}_j - \hat{S}_j^2\right), \\ \hat{\tau}_i &= \frac{\hat{S}_i(v_i - \hat{q}) - \hat{Q}_i(\hat{p} - \hat{q})}{Y_i} \\ &\quad - \frac{(\overline{Y} - Y_i)}{(Y_1 + Y_2)Y_i} \sum_{i=1,2} \left(\hat{S}_i(v_i - \hat{q}) - \hat{Q}_i(\hat{p} - \hat{q})\right). \end{aligned} \tag{10}$$

4 Welfare Improving Strategies

The solution presented in the previous section, although an improvement on the "no cross border shopping" case, does not represent a welfare First Best because of the positive externality caused by the production of health care. In our model a

[7] See the Appendix for a formal proof.

coordinated solution that allows a First Best optimal allocation may be obtained as a "bottom up" solution where the two Regions freely decide for a coordinated policy where quantities and mobility are set through a bargaining process or for a top down procedure where the supernational authority decides how much to produce and where. This solution can be attained only if specific conditions are met; in order to describe the process in the simplest way, we will start by presenting the optimal solution that in both cases derives from the maximization of the following welfare function:

$$\begin{array}{c} \max_{p_1,p_2,S_1,S_2} W^* = W_1(p_1,S_1,S_2) + W_2(p_2,S_2,S_1) \\ \text{s.t.} \\ \tau_i = \frac{v_iS_i+q(Q_i-S_i)-G_i}{Y_i}, \\ S_1+S_2 = Q_1+Q_2 = Q. \end{array} \tag{11}$$

The solution is derived in the Appendix and can be written as:

$$\begin{aligned} p_i &= q, \\ S_i &= \frac{1}{2}\frac{q - v_i + z_iw_i + z_j(1-w_j)}{w_i+1-w_j}, \quad j \neq i \in \{1,2\}, \\ q &= \beta\frac{(w_1-w_2)^2+(z_2w_2-z_1w_1)(w_1-w_2)-1}{(w_1-w_2)^2-(1+\beta)} \\ &+\frac{\beta}{2}\frac{(z_2-v_1)(w_2+(1-w_1))+(z_1-v_2)(w_1+(1-w_2))}{(w_1-w_2)^2-(1+\beta)} \end{aligned} \tag{12}$$

which can be interpreted as follows: the allocation of production between the two local authorities should follow an efficiency principle by balancing the need to reduce the cost of public provision with the utility both communities derive from the location of the production of that specific good.

In general $q^* < \hat{q}$ because in this case the quantity to be supplied takes into full account the effect of the spillover produced by the supply of health care. Production will be concentrated in the more productive Region, but in this case the full effect of the reduction in welfare in the sending Region is taken into account.

The quantity of the impure public good in the two local authorities is the same in equilibrium. p^* is chosen to equalize the marginal rate of substitution between income and the impure public good with the price ratio.

The optimal solution in terms of p^*, S_1^*, S_2^*, Q_1^* and Q_2^* can be substituted in the welfare function to obtain:

$$\begin{aligned} W^* &= Y_i(1-\tau_i^*) + \frac{\beta^2 - p_i^{*2}}{4\beta} + w_i(z_iS_i^* - S_i^{*2}) + (1-w_i)\left(z_iS_j^* - S_j^{*2}\right), \\ \hat{\tau}_i &= \frac{S_i^*(v_i-q^*) - Q_i^*(p^*-q^*)}{Y_i} - \frac{(\overline{Y}-Y_i)}{(Y_1+Y_2)Y_i}\sum_{i=1,2}(S_i^*(v_iq^*) - Q_i^*(p^*-q^*)). \end{aligned} \tag{13}$$

Total welfare is clearly increasing in this equilibrium which represents First Best. However, this does not necessarily mean that both local authorities are better off. In fact the reimbursement price for the health care supplied across the border is lower than in the previous case and although the total quantity (hence the utility derived from the public characteristic of health care) increases, the level of taxation in the destination Region will increase. The second element that determines the difference in welfare is the equalization grant. The quantity of health care produced is increasing, the average cost is lower than in the decentralized case, but total expenditure is increasing. This implies that the equalization grant is increasing as well. In general this implies that more resources are flowing from the rich to the poor Region. If the rich Region is also the more productive one (i.e. the one receiving the flow of patients), both elements (cross border shopping and coordination) will have a depressing effect on local welfare. On the other hand, if the poor Region is the more efficient, the equalization grant will be able to partially compensate for the increased taxation due to cross border health care provision.

5 Discussion and Numerical Example

The model presented in this chapter is a first attempt to study the welfare properties of opening international trade in the health care sector. We study the problem from the perspective of local or national governments where health care is supplied by the public sector and has the double characteristics of being a paternalistic good and a local public good with spillovers. The first characteristics means that the good will be supplied for free at the point of use, and its cost will be financed through a linear income tax. The Government rations the quantity of health care available by determining the marginal individual that is allowed to receive care. The latter depends on the utility derived from health care, on its production cost. However, utility has two components: the benefit received by the individual in terms of accrued health and a public good element that depends on the quantity of care produced in each Region. This implies that the benefit of the marginal patients will always be lower than the production cost of health care. When cross border supply is not allowed, the only effect we observe is a redistribution of income from non users of the service to users. In fact the former pay taxes that will then be used to finance health care. When cross border shopping is allowed this redistribution effect may spread to the regional level.[8]

International trade allows the less efficient Region to buy health care in the more efficient one; in this way the number of people allowed to receive health care increases, and its average production cost decreases. Although health care

---

[8] The optimal conditions for the provision of an impure public good imply that the production cost is equal to the marginal private benefit increased by the utility deriving from the public good nature of the good. See Musgrave and Musgrave (1989) for a formal proof.

production increases, the quantity of health care locally produced decreases; this implies that utility from the public good is likely to decrease. In general, these effects produce an increase in welfare. On the other hand, the more efficient Region may not necessarily gain in this process. Fewer resident patients may be allowed to receive health services owing to an increase in the threshold to be eligible (which optimally equates q, the price that clears the market). More health care is produced in the efficient Region, which implies an increase in welfare, but the price (q) for cross border shopping is going to be lower than the marginal cost, i.e. the tax rate will have to increase. The equalization grant has a countervailing effect in this case: the total cost to produce health care may decrease and the tax rate in the more efficient local authority certainly increases.

In general we can state that when two Regions cooperate, total welfare increases because the spillovers effects are correctly taken into account, but the gain may be unequally split between the two.

The analytic analysis of this case is quite cumbersome. Below we present some simulations that provide some numerical insights into the working of our model. The general form of the welfare function is:

$$W_i = Y_i(1-\tau) + \frac{\beta^2 - p_i^2}{4\beta} + w_i\left(z_i S_i - S_i^2\right) + (1-w_i)\left(z_i S_j - S_j^2\right).$$

We have evaluated the different solutions for the following initial parameters $Y_1 = 1.5, Y_2 = 1, w_1 = 0.65, w_2 = 0.65, v_1 = 5, v_2 = 6, \beta = 20, z_1 = 1.5, z_2 = 1.5$. The solution has been computed for two systems, with and without an expenditure-based equalization grant. The latter solution should correspond to a bilateral agreement between two national States while the first one is more likely to represent the case of an upper tier which is also interested in local income redistribution.

The results are presented in Table 1. Welfare reaches its maximum in a system where cross border shopping is allowed and the level of care produced is jointly determined, as one might expect. In this case production is concentrated in local authority 1, but the quantity demanded is the same in both local authorities. The tax rate and the grant depends on q, the price for mobility, as much as welfare of the two local authorities. The decentralized solution with cross border shopping is characterised by a higher price for the quantity traded, which creates an increase in the quantity of the local good produced in 1, but a reduction in total demand $Q_1 + Q_1$. This solution is characterised by the lowest average tax rate which implies that the equalization grant is minimum.

The welfare comparisons for this case are quite interesting. Let us start with the case where an equalization grant is present. For Region 1 opening to trade is not welfare improving. This is due to the combined effects of several elements: from the expenditure side we note that although total quantity and local production increase, health care available to residents decreases; on the financial side, local taxes increase because the reduction in the cost to produce health care is mainly borne by Region 1. The welfare loss persists also in the case where an equalization grant is not foreseen. This means that in this case the welfare loss is mainly due to

Table 1 Simulation results

	No mobility	Mobility	First best
q		0.86854	0.68182
S_1		0.64888	0.59091
S_2		0.26427	0.34091
Q_1	0.46831	0.45657	0.46591
Q_2	0.44484	0.45657	0.46591
Q	0.91315	0.91315	0.93182
p_1	0.63380	0.86854	0.68182
p_2	1.10330	0.86854	0.68182
Bilateral agreement			
τ_1	0.31221	0.32124	0.33712
τ_2	0.66725	0.56343	0.59659
G_1	0	0	0
G_2	0	0	0
W_1	4	3.9726	3.9702
W_2	3.2765	3.3233	3.3366
Upper tier decision			
τ_1	0.38791	0.39092	0.41061
τ_2	0.55370	0.45890	0.48636
G_1	−0.11356	−0.10453	−0.11023
G_2	0.11356	0.10453	0.11023
W_1	3.8864	3.8680	3.8599
W_2	3.3901	3.4278	3.4469
$W_1 + W_2$	7.2765	7.2958	7.3068

The values of the parameters are: Y_1 and Y_2 are independent and uniformly distributed. Their average is equal to 1.5 and 1 respectively. $w_1 = 0.65$, $w_2 = 0.65$, $v_1 = 1$, $v_2 = 1.5$, $\beta = 10$, $z_1 = 1.5$, $z_2 = 1.5$

the effect of q on local tax rates. More coordination further reduces the welfare of Region 1, especially in the presence of an equalization grant. This opens an interesting policy debate about cross border supply and federalism. Unless the good is perceived as a local public good, this policy is imposed by an upper level, it is not likely to be the outcome of coordination between local authorities. This may partially explain the present debate at EU level about patient mobility across countries (Glinos et al. 2010; Brekke et al. 2011; Legido-Quigley et al. 2012). Most of the agreements that have been put forth are related to planned admissions, a type of activity where the public good aspect of health care is probably less important as well as the spillover effect. In these cases agreements may be feasible and are beneficial to both Regions because the price for non residents is going to be quite close (if not equal) to the marginal cost of production.

In order to study the effects of the cross border price on welfare, in Table 2 we present a second simulation. The parameters are the same, but in this case Region 2 (the poorer) is the most efficient. Also in this case the most efficient Region is the loser in terms of welfare, but now the equalization grant has a mitigating effect on

Table 2 Simulation results

	No mobility	Mobility	First best
q		0.86854	0.68182
S_1		0.64888	0.59091
S_2		0.26427	0.34091
Q_1	0.44484	0.45657	0.46591
Q_2	0.46831	0.45657	0.46591
Q	0.91315	0.91315	0.93182
p_1	1.10330	0.86854	0.68182
p_2	0.63380	0.86854	0.68182
Bilateral bargain			
τ_1	0.44840	0.32124	0.39773
τ_2	0.46831	0.56343	0.50568
G_1	0	0	0
G_2	0	0	0
W_1	3.7765	3.9726	3.8366
W_2	3.5000	3.3233	3.4702
Upper tier decision			
τ_1	0.52054	0.4453	0.47121
τ_2	0.35475	0.37732	0.39545
G_1	−0.11356	−0.10453	−0.11023
G_2	0.11356	0.10453	0.11023
W_1	3.6630	3.7187	3.7264
W_2	3.6135	3.5771	3.5804
$W_1 + W_2$	7.2765	7.2958	7.3068

The values of the parameters are: Y_1 and Y_2 are independent and uniformly distributed with average 1.5 and 1 respectively. $w_1 = 0.65$, $w_2 = 0.65$, $v_1 = 1$, $v_2 = 1.5$, $\beta = 10$, $z_1 = 1.5$, $z_2 = 1.5$

the loss, as one might expect. It is also interesting to note that in this case more coordination is preferred to a simple trade agreement.

These examples show that it is not possible to determine the effects of international trade on welfare distribution. Total welfare will certainly increase, but international trade might not be a Pareto superior solution. The distribution of the welfare between the two Regions depends on several factors: the importance of spillovers (measured by w), the relative importance of the public good aspect of health care (measured by z) and, if an equalization grant is foreseen, the difference in income in the two Regions.

To show this in our model, we have determined the value of w (the spillover parameter) for which trade and cooperation is a Pareto efficient solution. For the first simulation (Region 1 richer and more efficient) trade would be welfare improving for both Regions for $w = 0.79$; if $w > 0.85$ even coordination improves welfare. These values are lower (0.72 and 0.81, respectively) without equalization grant.

For the second simulation (Region 1 richer and Region 2 more efficient) the value of w is equal to 0.81 for trade agreement and to 0.79 for a coordinated solution. These values are however higher without an equalization grant.

Finally, if health care is perceived mainly as a private good, international trade is more likely to have positive effects. In this case, in fact, the price q for cross border supply comes closer to the marginal of the more efficient Region. This Region will then experience a (relatively) lower increase in the local tax rate. This may explain why some countries have set agreements for cross border supply of non acute hospital care. These services are quite similar to private goods and, in this case, international trade may be a Pareto superior solution.

6 Conclusions

The development of trade and international relations has put increasing pressure on regulatory authorities to define new rules for health care cross border shopping. The process of globalization and the introduction of new technologies means that the market for health care is becoming more and more international. Telemedicine and telediagnosis allow patients to be treated and monitored at an increasing distance, often beyond the national boundaries. Fidler et al. (2009) argue that while for international trade rules are well defined by GATT and other treaties, for health care settlements they are not so clear and only about 40 % of international trade is carried out under these rules. International trade challenges the governance of health care systems in any country, but it is going to put extra pressure on public health care systems where the access to health care needs to be strictly regulated. In a global context these restrictions may be more difficult to enforce and may be challenged on legal grounds.[9]

In Europe economic integration has meant that also health care systems have to face the challenge of cross border shopping, both at demand and supply level. Although in Europe health care expenditure related to cross border shopping is relatively small, the number of people traveling abroad to receive care is increasing and shows a positive trend. The EU has started regulating these flows (Greer et al. 2013), but very little is known on the economic impact of such mobility.[10]

In this chapter we show that cross border shopping and coordinated efforts among Regions may not be compatible with local welfare maximization. This is because health care may be often used to redistribute income both among individuals and among Regions. If this is the case, cross border shopping allows reduction of the total cost of health care produced, but it may allocate the benefits of such reduction in a very uneven way. This is due to two different effects: (i) the presence of an equalisation grant and (ii) the spillover effect produced by health care.

---

[9] For example, the restrictions on the number of providers that can deliver health care is a controversial matter because this rule introduces restrictions to international trade.

[10] For a discussion of the open issues see Mackenbach et al. (2013).

This second element is particularly important because its role has been often overlooked. The public good aspect of health care implies that both the internal and the international price for health care are lower than the production cost. This means that if a country receives patients from outside, the application of the First Best rules may imply that the price (q) may be well below the production cost. All else being equal, the local authority receiving patients may experience an increase in its tax rate. Its utility will increase since utility derived from the option good aspect of health care is increasing, but this might not be enough to compensate for the increased tax rate.

In our model, mobility improves welfare for both Regions only if health care is a paternalistic good and there are no equalization grants. In all the other cases, the effect may be ambiguous and this means that a coordinated solution where cross border mobility is allowed may not be the outcome of this game. Our results allow us to explain why lower Government tiers (both within the same country or in the context of international trade agreements) may be reluctant to allow their patients to travel abroad to receive health care. The problem is particularly important in countries where income is unevenly distributed across jurisdictions so that the equalization grant plays a very important role in financing expenditure.

The EU directive on cross border mobility is at present foreseeing a reimbursement based on the cost to produce the service. This criterion represents a more favorable arrangement for the more efficient Region that will not have to subsidise the services produced for non residents, but it might not represent a First Best solution in a context where health care is an impure public good that produces spillovers effects on other Regions.

Appendix

Solution to the Problem Without Cross Border Supply

When mobility is not allowed, the quantity produced in each Region is equal to demand; in this case it is sufficient to find the utility for the marginal consumer allowed to use health care. The problem can be written as:

$$\max_{p_i} W_i(p_i, Q_i, Q_j)$$
$$\text{s.t.}$$
$$\tau_i = \frac{v_i Q_i - G_i}{Y_i},$$
$$Q_i = \frac{\beta - p_i}{2\beta}.$$

The constraints can be substituted back into the maximization problem that can be solved as an unconstrained maximization:

$$\max_{p_i} \overline{W}_i = Y_i - v_i\left(\frac{\beta - p_i}{2\beta}\right) + G_i + \frac{\beta^2 - p_i^2}{4\beta}$$
$$+ w_i\left(z_i\frac{\beta - p_i}{2\beta} - \left(\frac{\beta - p_i}{2\beta}\right)^2\right) + (1 - w_i)\left(z_i\frac{\beta - p_j}{2\beta} - \left(\frac{\beta - p_j}{2\beta}\right)^2\right).$$

The FOC can be written as:

$$\frac{\partial \overline{W}_i}{\partial p_i} = 0 \Rightarrow \frac{v_i - p_i}{2\beta} + \frac{\beta w_i(1 - z_i) - p_i w_i}{4\beta^2} = 0,$$

which is solved for the value $\overline{p}_i$ in the text.

Solution to the Problem with Cross Border Supply

When mobility is allowed, the quantity produced in each Region may not be equal to demand; in this case each Region maximizes for p_i and S_i, but in a subsequent step they will have to define a price for mobility and the number of patients allowed to cross the border. The problem can be written as:

$$\max_{p_i, S_i} W_i(p_i, S_i, S_j)$$
$$\text{s.t.}$$
$$\tau_i = \frac{v_i S_i + q(Q_i - S_i) - G_i}{Y_i},$$
$$Q_i = \frac{\beta - p_i}{2\beta}.$$

The constraint can be substituted back into the maximization problem that can be solved as an unconstrained maximization:

$$\max_{p_i, S_i} \hat{W}_i = Y_1\left(1 - \frac{v_i S_i + q(\frac{\beta - p_i}{2\beta} - S_i) - G_i}{Y_i}\right) + \frac{\beta^2 - p_i^2}{4\beta}$$
$$+ w_i(z_i S_i - S_i^2) + (1 - w_i)\left(z_i S_j - S_j^2\right).$$

The FOCs can be written as:

$$\frac{\partial \hat{W}_i}{\partial p_i} = 0 \Rightarrow \frac{q - p_i}{2\beta} = 0,$$
$$\frac{\partial \hat{W}_i}{\partial S_i} = 0 \Rightarrow -v_i + q + w_i z_i - 2w_i S_i = 0,$$

which allow to obtain the solutions in the text in terms of q. The second step consists of reconciling demand with supply as explained in the text.

Solution to the Central Government Problem

The more general problem can be written as:

$$\max_{p_i, S_i, q} W_1(p_1, S_1, S_2) + W_2(p_2, S_2, S_1)$$

s.t.

$$\tau_i = \frac{v_i S_i + q(Q_i - S_i) - G_i}{Y_i},$$
$$Q_i = \frac{\beta - p_i}{2\beta},$$
$$Q = S_1 + S_2 = Q_1 + Q_2.$$

The first and the second constraints can be substituted in the maximization problem while the third one will be used to define the Lagrangian for the problem

$$\max_{p_i, S_i, q} \mathcal{L} = \sum_{i \neq j = 1,2} Y_i \left(1 - \frac{v_1 S_1 + q(\frac{\beta - p_i}{2\beta} - S_i) - G_i}{Y_1}\right) + \frac{\beta^2 - p_i^2}{4\beta}$$
$$+ w_i(z_i S_i - S_i^2) + (1 - w_i)\left(z_i S_j - S_j^2\right) - \lambda\left(\sum_{i=1,2} 1 - \frac{\beta - p_i}{2\beta} - \sum_{i=1,2} S_i\right).$$

The FOC for the problem can be written as:

$$\frac{\partial \mathcal{L}}{\partial p_1} = 0 \Rightarrow -\frac{1}{2}\frac{-q + p_i - \lambda}{\beta} = 0,$$
$$\frac{\partial \mathcal{L}}{\partial S_i} = 0 \Rightarrow z_i w_i + z_j(1 - w_j) - 2S_i(1 + w_i - w_j) - v_i + q + \lambda = 0,$$
$$\frac{\partial \mathcal{L}}{\partial \lambda} = 0 \Rightarrow \sum_{i=1,2} 1 - \frac{\beta - p_i}{2\beta} - \sum_{i=1,2} S_i = 0,$$
$$\frac{\partial \mathcal{L}}{\partial q} = 0 \Rightarrow \sum_{i=1,2} Q_i - \sum_{i=1,2} S_i = 0.$$

In this form the solution is not determined because the derivative for q is not defined.

However, let us now observe the constraint: if it is satisfied ($\lambda = 0$), the market is in equilibrium, i.e. the demand for health care is exactly equal to the supply of health care. In any other case rationing exists either on the demand or on the supply. Using the same procedure described in Sect. 3.2 we can find q that clears the market by adding the constraint $\lambda = 0$ to the problem described in the FOCs above. The q that clears the market is the solution of the following problem:

$$1-\frac{q}{\beta}-\frac{1}{2}\frac{q-v_1+w_1z_1+z_2(1-w_2)}{w_1+1-w_2}-\frac{1}{2}\frac{q-v_2+w_2z_2+z_1(1-w_1)}{w_2+1-w_1}.$$

Comparisons

Quantity of Health Care

In order to determine the quantity of health care produced, we can start by evaluating the difference in the marginal private benefit of health care in the different model which is represented by p.

Difference Between First Best and Cross Border Supply

In this case we know from the theory (Koethenbuerger 2008; Oates 2008) that the quantity in First Best increases. To show this, let us consider the difference between $\hat{q}$ and q^*which can be written as:

$$\begin{aligned}
&\frac{w_2(v_1+w_1(1-z_1))+w_1(v_2+w_2(1-z_2))}{w_2(\beta+w_1)+w_1(\beta+w_2)}\beta\\
&-\frac{1}{2}\beta\frac{(w_i-w_j-1)v_i-(w_i-w_j+1)v_j+2(w_i-w_j)^2}{-1-\beta+(w_i-w_j)^2}\\
&-\frac{1}{2}\beta\frac{\left(z_i+z_j-2w_i^2z_i-2w_j^2z_j+(w_i-w_j)(z_i-z_j)+2w_iw_j(z_i+z_j)-2\right)}{-1-\beta+(w_i-w_j)^2}
\end{aligned}$$

and, for $w_i=w_j=1$

$$\frac{v_i+2-z_i+v_j-z_j}{2\beta+2}\beta-\frac{\beta}{2}\frac{-v_i-v_j-2+z_i+z_j}{-1-\beta}=0,$$

as expected.

Total quantity Q is clearly bigger in FB than in the cross border supply equilibrium.

Difference Between the "No Cross Border Supply" and the Case When Cross Border Supply Is Allowed

In this case it can be written as:

$$\begin{aligned}
p_i-q=\beta&\frac{v_i+w_i(1-z_i)}{\beta+w_i}\\
&-\beta\frac{w_j(v_i+w_i(1-z_i))+w_i(v_j+w_j(1-z_j))}{w_j(\beta+w_i)+w_i(\beta+w_j)}.
\end{aligned}$$

Let us start by assuming that $w_i=w_j=w$. In this case the difference can be written as:

$$p_i - q = \beta \frac{v_i + w(1 - z_i)}{\beta + w} - \frac{1}{2} \frac{(v_i + w(1 - z_i)) + (v_j + w(1 - z_j))}{(\beta + w)} \beta.$$

The two expressions are equal for

$$z_j = z_i + \frac{v_j - v_i}{w}.$$

In this case total quantity depends on the interactions of the parameters. The difference can be written as:

$$\bar{Q} - \widehat{Q} = 1 - \frac{1}{2}\left(\frac{v_i + w_i(1 - z_i)}{\beta + w_i} - \frac{v_j + w_j(1 - z_j)}{\beta + w_j}\right) - \left[1 - \frac{w_j(v_i + w_i(1 - z_i)) + w_i(v_j + w_j(1 - z_j))}{w_j(\beta + w_i) + w_i(\beta + w_j)}\right].$$

For $w_i = w_j = w$ total quantity is the same, i.e. $\bar{Q} = \widehat{Q}$.

References

Brekke, K. R., Levaggi, R., Siciliani, L., & Straume, O. R. (2011). Patient mobility, health care quality and welfare. CEPR Discussion Papers 8559, C.E.P.R. Discussion Papers.

Dahlby, B., & Wilson, L. S. (1994). Fiscal capacity, tax effort, and optimal equalization grants. *Canadian Journal of Economics, 27*(3), 657–672.

De la Fuente, A., Doménech, R., & Rant, V. (2010). Addressing the net balances problem as a prerequisite for eu budget reform: A proposal. *CESifo Economic Studies, 56*(2), 221–250.

Fidler, D. P., Drager, N., & Lee, K. (2009). Managing the pursuit of health and wealth: The key challenges. *Lancet, 373*(9660), 325–331.

Frankel, J., & Romer, D. (1999). Does trade cause convergence. *American Economic Review, 89*, 379–399.

Glinos, I., Baeten, R., & Maarse, H. (2010). Purchasing health services abroad: Practices of cross-border contracting and patient mobility in six European countries. *Health Policy, 95*(2), 103–112.

Greer, S. L., Hervey, T. K., Mackenbach, J. P., & McKee, M. (2013). Health law and policy in the European Union. *The Lancet, 381*(9872), 1135–1144.

Koethenbuerger, M. (2008). Revisiting the "decentralization theorem"—On the role of externalities. *Journal of Urban Economics, 64*(1), 116–122.

Legido-Quigley, H., Glinos, I., Baeten, R., & McKee, M. (2007). Patient mobility in the European Union. *British Medical Journal, 334*(7586), 188–190.

Legido-Quigley, H., Glinos, I. A., Baeten, R., McKee, M., & Busse, R. (2012). Analysing arrangements for cross-border mobility of patients in the European Union: A proposal for a framework. *Health Policy, 108*(1), 27–36.

Levaggi, R. (2010). From local to global public goods: How should we write the utility function. *Economic Modelling, 27*(5), 1040–1042.

Levaggi, L., & Levaggi, R. (2014). Patients mobility across borders: A welfare analysis. In Levaggi, R., & Montefiori, M. (Eds.). *Health care provision and patient mobility: Health integration in the European Union*. Berlin: Springer.

Lindsay, C. M. (1969). Option demand and consumer's surplus. *The Quarterly Journal of Economics, 83*(2), 344–346.

Mackenbach, J. P., Karanikolos, M., & McKee, M. (2013). The unequal health of Europeans: Successes and failures of policies. *Lancet, 381*(9872), 1125–1134.

Musgrave, R. A., & Musgrave, P. B. (1989). *Public finance in theory and practice*. Finance series. Singapore: McGraw-Hill.

Oates, W. E. (2008). On the evolution of fiscal federalism: Theory and institutions. *National Tax Journal, 61*, 313–334.

Petretto, A. (2000). On the cost-benefit of the regionalisation of the national health service. *Economics of Governance, 1*, 213–232. doi:10.1007/PL00011014.

Schnellenbach, J. (2012). Nudges and norms: On the political economy of soft paternalism. *European Journal of Political Economy, 28*(2), 266–277.

Smart, M. (1998). Taxation and deadweight loss in a system of intergovernmental transfer. *The Canadian Journal of Economics, 31*(1), 189–206.

Smith, R. D., Chanda, R., & Tangcharoensathien, V. (2009). Trade in health-related services. *Lancet, 373*(9663), 593–601.

Tommasi, M., & Weinschelbaum, F. (2007). Centralization vs. decentralization: A principal-agent analysis. *Journal of Public Economic Theory, 9*(2), 369–389.

MIX
Papier aus verantwortungsvollen Quellen
Paper from responsible sources
FSC® C105338

If you have any concerns about our products,
you can contact us on
ProductSafety@springernature.com

In case Publisher is established outside the EU,
the EU authorized representative is:
Springer Nature Customer Service Center GmbH
Europaplatz 3, 69115 Heidelberg, Germany

Printed by Libri Plureos GmbH
in Hamburg, Germany